t-reversal p. 270 p. 203 Supersymmetry hypercube

$x \to (y \to z)$ p. 247

$\equiv (x \to y) \to (x \to z)$

MATRIX LOGIC AND MIND

A probe into a unified theory of mind and matter

MATRIX LOGIC AND MIND

A probe into a unified theory of mind and matter

AUGUST STERN

Intelligent Systems Technology Europe
Amsterdam, The Netherlands

1992

NORTH-HOLLAND / ELSEVIER
AMSTERDAM • LONDON • NEW YORK • TOKYO

ELSEVIER SCIENCE PUBLISHERS B.V.
Sara Burgerhartstraat 25
P.O. Box 211, 1000 AE Amsterdam, The Netherlands

Distributors for the United States and Canada:

ELSEVIER SCIENCE PUBLISHING COMPANY INC.
655 Avenue of the Americas
New York, N.Y. 10010, U.S.A.

ISBN: 0 444 88798 9

Printed in The Netherlands

To my country,
Holland

Thefundamentalideasofscienceareessen
tiallysimpleandmayasarulebeexpressed
inalanguagecomprehensibletoeveryone

Albert Einstein

ACKNOWLEDGEMENTS

I wish to thank Prof. J.L. Mackenzie of Amsterdam for improving the quality of the text, Dr A. Colter of Los Alamos for his extensive comments, H. Labega for assistance with the graphics and Dr H. Aertsen for his help with the computer.
The editorial advice and expertise of Drs A. Sevenster of Elsevier Science Publishers was indispensable.

Amsterdam August Stern

CONTENTS

Introduction 1

Chapter 1 Conventional logic 5

1.1 Deterministic systems 6

1.1.1 Binary intelligence 6

1.1.2 Multivalued logic 26

1.2 Probability logic 29

1.2.1 Negation 30

1.2.2 Conjunction 31

1.2.3 Disjunction 33

1.2.4 Conditional 35

Chapter 2 Matrix logic 37

2.1 Logic space 37

2.2 Discrete universe $\mathbf{\Omega}^{\square}$ 46

2.2.1 Atomic transformations 47

2.2.2 Matrix strings 79

2.2.3 Intelligence code 89

2.3 Continuous universe $\mathbf{\Omega}^{\lozenge}$ 97

2.3.1 Semitruth 97

2.3.2 Multiplication table 105

2.3.3 Correction terms 109

2.3.4 Universum operator 111

2.3.5 Differential logic 122

Chapter 3 Intelligent interactions 137

3.1 Antilogic 137

3.1.1 Matrix inverses 138

3.1.2 Vacuum interactions 151

3.1.3 Chronologic 156

3.2 Artificial intelligence 160

3.2.1 Forbidden interactions 161

3.2.2 Autoproducts 164

3.3 Logic quantum numbers 173

3.3.1 Observables 174

3.3.2 Logic eigenequations 176

3.3.3 Compatibility 187

3.3.4 Connectives as numbers 189

3.3.5 Diagonalization 193

3.3.6 Fractal quantum mechanics 195

3.4 Annihilation and creation operators 197

3.4.1 Spacetime diagrams 197

3.4.2 Concept formation 205

3.4.3 Bosons and fermions 206

3.4.4 A logic model of matter 215

3.5 Tensor logic 229

3.5.1 Hyperconnectives 229

3.5.2 Fractal logic operators 234

3.5.3 Logic subspaces 236

Chapter 4 Problems and solutions 241

Index 287

NOTATIONS

Universes

Ω^+	–	positive logic
Ω^-	–	negative logic
$\Omega^\square$	–	discrete logic
$\Omega^\Diamond$	–	modal logic
Ω^n	–	tensor logic

Operators

L	–	logic operator
H	–	hyperspace operator
F	–	fractal operator
Ω	–	universum operator

Operations

	–	matrix multiplication
$\sim$	–	matrix transposition
$\sqcap$	–	converse transposition
✹	–	transmultiplication
$\otimes$	–	tensor multiplication
[,]	–	commutator
θ	–	orthogonal transform
Tr	–	trace

INTRODUCTION

Mathematical description is essential to the scientific method. Whereas classical logic was devised mainly to articulate and communicate logical truths, the use of symbolic notation has brought logic closer to the realm of mathematics. However, in spite of major advances symbolic logic has not yet attained a genuinely computational status. Though the boundary between logic and mathematics is somewhat vague, it is nevertheless real and significant.

The primary goals of this work are to close the gap between logic and mathematics and to bring logic fully into the domain of the exact sciences. Introducing the concept of logic space and interpreting logic connectives as matrix operators acting in the two adjoint spaces of logic vectors, we develop the fundamental operator formulation of logic. The discovery of links between logic and vector spaces leads to a coherent theory, in which the power of direct mathematical computation is for the first time applied to the complete set of logic operations. The application of the powerful mathematical apparatus of matrix algebra allows us to describe logic transformations as operations of symmetry in logic space. The formalism that is developed displays new deductive and inductive power, since it not only allows us to derive the classical results of conventional logic but also permits their generalization, making available new techniques of logical inference. Applicable in a wide variety of fields, matrix logic demonstrates a great potential, making it a useful tool with a broad spectrum of theoretical and technical applications in computer science, fundamental physics, as well as in mathematics and logic themselves.

It is necessary to emphasize that matrix logic does not reject conventional logic. The significance of the new formalism resides in the fact that, in addition to generating new important results, it also incorporates in itself conventional logic as a special scalar limit. Therefore, conventional logic can be formulated, if necessary, in the form of matrix logic equations. The converse, however, is not universally applicable.

The study of logic has hitherto been carried out by developing and investigating different forms of scalar logic. However, deciphering the puzzles of the mind requires not simply a modification of scalar logic but an entirely new generalization, in the same way as in physics we describe different phenomena with clearly distinct quantities such as scalars, vectors, tensors and others. The important innovation that is introduced in matrix logic is that we place at the foundation of logic not scalar values but more complex mathematical objects, namely logic vectors and logic operators, joined eventually into the more general concept of a logic tensor. Following this line of reasoning we begin to understand that the field of logical truth is much wider and the structure of logic connectives much more complex than was previously recognized.

The concept of logic space is instrumental in revealing the tensor nature of logic quantities: the logic scalars are obtained as the inner product of logic vectors, and the logic operators as the outer product of the same logic vectors. Consequently, the logic vectors alone become sufficient for the construction of the logic theory. We obtain conventional logic by scalarization of matrix logic, whereas vectorization or more generally tensorization of logic values is required for the reverse transition:

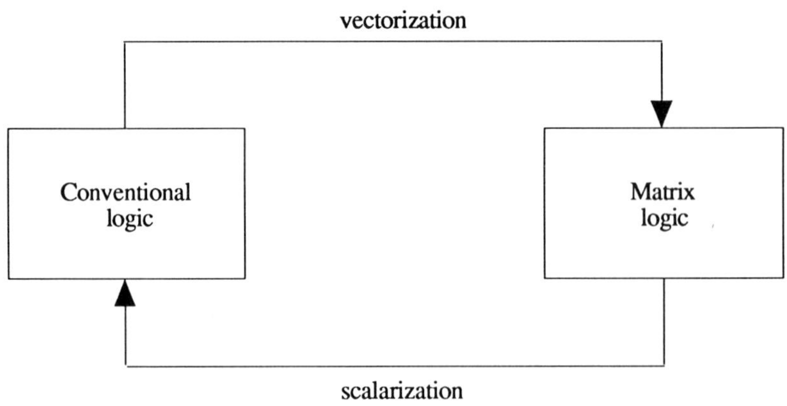

With a wide range of computational capabilities, matrix logic solves problems inaccessible to other forms of logic. It allows us to study logic connective operators in isolation, just as differential or integral operators are examined in mathematics and theoretical physics.

One of the fundamental results of marix logic is the possibility of the direct interaction of logic connectives. Inconceivable in conventional logic, the interaction of connectives introduces a higher level of abstraction.

With the logic operations becoming genuinely mathematical operations, we extend logic into the domain of modal continuous values. Furthermore, exploring the potential of the matrix operator formulation, we are able to derive both discrete and modal logics from the same formal base!

A conspicuous result of continuous matrix logic is a significant enlargement of the set of logic connective operators. There is a trade-off between a weakening of the integer logic values with the transition from a discrete to a continuous universe and an increase in the operator power of logic, with a concomitant decrease in the number of forbidden logic interactions.

Applying the concept of the matrix inverse to logic operators, we further extend logic into the domain of negative logic antivalues and associate them with the logic of antimatter in the direct quantum-relativistic sense. As a consequence of these different extensions of logic we establish the fundamental quaternary alphabet of truth-values

$$E_4 = \{-1, 0, 1, 2\}$$

as compared with the two binary truth-values of conventional logic

$$E_2 = \{0, 1\} .$$

The truth-value 2 is extremely effective in formalizing sentences with double meaning, hence opening new avenues for overcoming the limitations of current machine translation technology. The negative truth-value −1 finds important applications in chronologic, an extension of logic which deals with logic operations realized forwards and backwards in time.
The matrix operator formulation of logic has not only greatly enhanced the computational power of logic, it also provides compelling reasons to view logic not as an abstract construct but as a fundamental structure underlying real physical interactions, which as such has to be included in the general system of the covariant laws of nature. This allows us to address logic valuations in terms of the spacetime diagram of quantum field theory. Since matrix logic permits logical processes to be defined in mathematical form similar or identical to the description of fundamental processes, we seek a synthesis of logical and physical methods in a unified theory whereby it is possible to achieve the logical description of physical processes and vice versa. The matrix logic method is so profoundly linked to the fundamental ideas of physics that it can only be properly understood in its relation to advanced physical theory.

The computational reform accomplished in the matrix operator formulation of logic achieves its ultimate significance in the new concept of logic quantum numbers. We show that the problem of logic and cognition in general can be formulated as the eigenvalue problem, analogous to the central problem of theoretical physics. Considering logic operators as observables, we obtain the set of logic eigenvalues

$$\lambda_i = \{-1, 0, 1, 2\} ,$$

which proves that the spectrum of logic operators is in exact correspondence with the fundamental alphabet of logic truth-values! This not only resolves the question of whether the operations of the mind are quantized, but also holds the key to their reduction to the universal and fundamental code of numbers. Since Aristotle, logic has come a long way from viewing logic connectives as abstract linguistic formations to identifying them as algebraic Boolean operations, then as matrix operators, and finally simply as numbers.

The theory of logic quantum numbers constitutes an important breakthrough in the study of the intelligence code which allows us for the first time to tackle the intractable problem of high-level intelligence in a scientific manner. We are compelled to conclude that the mechanism of cognition cannot be derived from either classical or quantum notions. A higher-order covariant theory is required in order to provide a plausible explanation for the fundamental effect of high-level intelligence.

With the fusion of physics and logic categories logic obtains the status of fundamental science. As a unified language which integrates a logical examination of the underlying phenomena of quantum theory and vice versa, matrix operator logic opens new avenues for the study of fundamental interactions and gives rise to the revolutionary conclusion that physics can be viewed and studied as a logic in a fundamental sense. We are on the verge of an unprecedented synthesis.

CHAPTER 1

CONVENTIONAL LOGIC

The birth of mathematical logic is associated with the work of George Boole, who formulated its initial principles and created the branch of mathematics known today as Boolean algebra. This development became possible upon the realization that if the simple propositions of logic could be represented by precise symbols, then the relation between them could be read as precisely as algebraic equations. Currently, logic theory that employs mathematical symbols is referred to as mathematical logic. However, conventional logic is more mathematical in presentation than in substance.

Mathematical logic, which initially appeared with the aim of dealing, in a concise symbolic code, with different aspects of deduction, has found important theoretical and practical applications far beyond its initial classical domain, that of the investigation of reasoning. At present mathematical logic continues to be developed, to a great extent under the influence of problems arising in applications.

Two complementary approaches can be distinguished in mathematical logic: deterministic and probabilistic. Traditional Boolean logic is deterministic in the sense that it restricts itself to the study of true and false logic expressions, with true and false representing the so-called binary alphabet of truth-values. A logic designed over such an alphabet is called a two-valued or binary logic. This convenient simplicity of the alphabet of truth-values, however, imposes severe limitations on the objects of logical inquiry, since it disregards all those expressions which cannot be defined in terms of true and false or as being definitely true or definitely false. The undecided, undetermined, unknown, or meaningless expressions encode either intermediate levels of validity or intermediate levels of certainty, and therefore their investigation requires multivalued or probabilistic methods.

The vocabulary of conventional logic is further expanded in modal logic, which deals with the concepts of necessity and possibility, and in quantum logic, which refutes the commutative and distributive laws of classical logic in order to accommodate the logical problems arising in quantum physics.

In this chapter we first consider the case of deterministic logic systems, followed by an overview of conventional probability logic.

1.1 Deterministic systems

1.1.1 Binary intelligence

Atomic and compound expressions

A distinction is made between elementary or atomic and compound logic expressions. Atomic expressions are denoted by small letters: x, y, z, ..., the compound expressions being denoted by capitals: X, Y, Z, ... The compound expressions are formed from atomic expressions, and like atomic expressions, they can also be identified as either true or false.
To form a compound expression in symbolic logic we introduce symbols for connectives. Traditionally, connectives are defined by so-called truth-tables, which specify the truth-values of the compound depending on the truth-values of the constituent atomic expressions.
The notion of connectives lies at the very foundation of logic. Together with atomic expressions, axioms and rules of inference, they actually represent the fundamentals from which a logic calculus can be designed.
Conventional logic distinguishes between unary and binary connectives. As the terminology indicates, the unary connective acts upon a single logic expression, either atomic or compound; the binary connective acts between two logic expressions, either atomic or compound.

Unary connectives

Affirmation or identity

It seems natural to begin the description of logic connectives with the identity connective, denoted by I. Affirmation of x is true if x is true and false if x is false. Thus the truth-table for the unary connective identity Ix is

x	Ix
f	f
t	t

where "t" both here and henceforth stands for "true", "f" standing for "false".

Negation

The unary connective negation, denoted by $\neg$, represents a fundamental logic operation. If x is an expression, $\neg$x (or $\overline{x}$) symbolizes the negation of x and is read "not x". The truth-table of negation is thus

x	$\neg$x
f	t
t	f

The negation of x is false if x is true , and true if x is false. Clearly, negation is the opposite of identity.

Logic constants

Two more unary connectives are defined in binary logic. These are the logic constants False and True, with the following truth-tables:

x	Fx
f	f
t	f

x	Tx
f	t
t	t

Independently of the truth-value of x, the logic constant in one case always obtains the truth-value f, in the other case the truth-value t.

Binary connectives

Two atomic expressions x and y may generate $2^2 = 4$ combinations of truth-values; ff, tf, ft, tt. Each combination set can yield $2^{2^2} = 16$ different truth-tables which also consist of the truth-values either true or false. With each of these 16 truth-tables conventional logic associates a corresponding binary Boolean connective.

Though, according to Jean Piaget, our intelligence makes use of all 16 logical operations, in most literature on logic only a few binary connectives are selected as theoretically and practically relevant: conjunction, disjunction, implication and equivalence. We now consider the truth-tables for this standard set of connectives.

Conjunction $\wedge$

The properties of conjunction appear to mirror some properties of the word "and": an expression of the form $x \wedge y$ is read as "x and y". The conjunction $x \wedge y$ is true if x and y are both true and is otherwise false. The truth-table for conjunction has the form

x	y	$x \wedge y$
f	f	f
f	t	f
t	f	f
t	t	t

Conjunction is a symmetric connective and the order of appearance of the atomic expressions is irrelevant for the result:

$$x \wedge y = y \wedge x .$$

We say x and y commute under conjunction.

Disjunction $\vee$

The disjunction of x and y is false if x and y are both false and is otherwise true. In ordinary language the connective $\vee$ corresponds to the word "or" which we use to separate two expressions. However, it is necessary to recognize that in language the disjunction "or" is applied in two different ways: it may mean that only x or only y, but not both, is true — the exclusive usage; it may also mean that either of them or both are true — the inclusive usage.
Sometimes exclusive disjunction is called strong alternation or XOR, while inclusive disjunction is called weak alternation. Disjunction is here understood in a weak sense and thus will be defined by the following truth-table:

x	y	$x \vee y$
f	f	f
f	t	t
t	f	t
t	t	t

It so happens that the use conventional logic makes of exclusive disjunction is negligible. Strong alternation is considered to be not as basic to our formal languages as weak alternation. In contrast, in matrix logic the role of strong alternation is indispensable.
Finally, note that disjunction, like conjunction, is a symmetric connective:

$$x \vee y = y \vee x .$$

We say x and y commute under disjunction.
It is necessary to emphasize that commutation of disjunction and commutation of conjunction hold only when we are concerned with the truth status of x and y and not with their meanings. Commutation, however, can fail to apply to causally dependent logical events occurring in time. In such cases commuting conjunction and disjunction must be replaced by sequential conjunction and sequential disjunction, where x and y are not permutable.

Implication →

Implication, "if x then y" or "x implies y", is denoted by x → y, where x is called the antecedent and y is called the consequent. By definition, implication is always true if the consequent is true or if the antecedent is false. This is quite different from the causal relation and one should not interpret logic implication in terms of cause and effect in a physical sense.

Though logic implication resembles the implication of ordinary discourse, it is nevertheless an imperfect approximation to the latter. But it does preserve the most important feature of ordinary reasoning: a true implication completely rules out the possibility of a true antecedent implying a false consequent. The truth-table of implication is thus defined as follows:

x	y	x→y
f	f	t
f	t	t
t	f	f
t	t	t

The truth-table for implication is a common source of confusion, which arises from not clearly separating the causal and logical aspects of implication. The difficulty arises when a false antecedent implies a true consequent. Since this can lead to meaningless statements, some people tend to think that such expressions are not useful. This is not so. Whereas the implication we are familiar with in common usage is based upon some connection in meaning between the antecedent and the consequent, a logic implication can represent only a relationship between the truth status of the antecedent and that of the consequent and not a relation between their meanings. In this context the significance of implication is that it enables the truth of y to be inferred from the truth of x, and nothing in particular to be inferred from the falsity of x.

Equivalence ↔

In common language the analogue of the logic connective equivalence is represented by "if and only if". In logic, equivalence $x \leftrightarrow y$ is true if and only if x and y have identical truth-values:

x	y	$x \leftrightarrow y$
f	f	t
f	t	f
t	f	f
t	t	t

The truth-table for equivalence presented above can be derived from the truth-tables of conjunction and implication since

$$(x \leftrightarrow y) = (x \rightarrow y) \wedge (y \rightarrow x)\,,$$

or from the truth-tables for conjunction, disjunction and negation since

$$(x \leftrightarrow y) = (x \wedge y) \vee (\overline{x} \wedge \overline{y})\,.$$

The above truth-tables for logic connectives represent an important operational base of symbolic logic. Without these simple, yet elegant constructs it would be quite difficult to imagine the successful development of mathematical logic we observe today.

Logic functions

Since the truth-value of a logic expression is fully determined by the truth-value of its constituent atomic expressions, every logic expression can be viewed as a certain Boolean function, the arguments of which, independently of each other, obtain the values "true" or "false". The same applies to the function itself. Because the arguments of logic functions and functions

themselves may take only two different values, there are a finite number of different argument sets for a function, and a logic function is uniquely defined by a table listing all the possible argument sets and corresponding function values.
It is convenient to define logic values in the binary number set

$$E_2 = \{0, 1\} .$$

by assigning to true expressions the numerical value 1, and to false expressions the numerical value 0.
Then any logic function $L(x_1, x_2, \ldots x_n)$ represents a mapping

$$\underbrace{E_2 \times E_2 \times \ldots \times E_2}_{\text{n times}} \rightarrow E_2 ,$$

with both the logic variables and logic function defined in the binary set E_2:

$$x_i \in E_2$$

$$L(x_1, x_2, \ldots x_n) \in E_2 .$$

The number of possible Boolean functions from n variables rapidly increases if n increases. For n = 1 there are 4 different functions, for n = 2, 16 functions, for n = 3 we can construct 256 functions, and for n = 4, 65536 functions.
If one wanted to find out whether some functions have some property, it would seem natural to try to scan the given set of functions, which we know in advance is finite. However, even for relatively small values of n, such a search becomes quite difficult even with the support of a computer.
Expressing compound Boolean functions via more simple functions, we can end up with functions which are considered as elementary Boolean functions, similar to sin x or x^2 considered as elementary in analysis. In the first section of this chapter we already gave a semi-formal description of some of these functions. Now we shall present the complete set of elementary Boolean functions and underline some of their properties relevant to our discussion.

Boolean functions of one argument

There are four Boolean functions of one argument: identity I(x), negation $\neg$(x), and the logic constants 1(x) and 0(x). The chart below shows that the structure of these functions is identical to the structure of the truth-tables for the unary connectives we considered above, except that the truth-values t and f are now replaced by the numerical values 1 and 0 respectively:

x \ L(x)	I(x)	⌉(x)	1(x)	0(x)
0	0	1	1	0
1	1	0	1	0

It is important to emphasize that the set of four Boolean functions of one argument is redundant; two of them can be expressed via the others:

$$I(x) = \rceil\rceil(x) \, ,$$
$$0(x) = \rceil 1(x) \, .$$

Consequently, only two of them, for example, the negation function and the logic constant 1(x), are needed to generate the rest of the set.

Boolean functions of two arguments

With two logic arguments x and y, we can construct sixteen different Boolean functions. These are presented explicitly in the following truth tables:

Equivalence ↔

x \ y	0	1
0	1	0
1	0	1

Inequivalence ≠

x \ y	0	1
0	0	1
1	1	0

Conjunction ∧

x \ y	0	1
0	0	0
1	0	1

Nand ↓

x \ y	0	1
0	1	1
1	1	0

Disjunction ∨

x \ y	0	1
0	0	1
1	1	1

Nor ↑

x \ y	0	1
0	1	0
1	0	0

Implication →

x \ y	0	1
0	1	1
1	0	1

Inhibition \

x \ y	0	1
0	0	0
1	1	0

Converse implication ←

x \ y	0	1
0	1	0
1	1	1

Converse inhibition /

x \ y	0	1
0	0	1
1	0	0

Filter ⊥

x \ y	0	1
0	0	0
1	1	1

Antifilter T

x \ y	0	1
0	1	1
1	0	0

Converse filter ⊣

x \ y	0	1
0	0	1
1	0	1

Converse antifilter ⊢

x \ y	0	1
0	1	0
1	1	0

Tautology constant $\hat{1}$

x \ y	0	1
0	1	1
1	1	1

Empty operator $\hat{0}$

x \ y	0	1
0	0	0
1	0	0

The logic functions we consider are characterized by being applicable to two arguments in a certain order. With each logic function we associate a converse function, in which the order of arguments is reversed. A logic function for which

$$L(x,y) = L(y,x)$$

is called symmetric; otherwise it is asymmetric.

Under these criteria the set of sixteen logic functions is divided into an octet of symmetric and an octet of asymmetric functions:

$$\left\{ \leftrightarrow, \neq, \wedge, \vee, \uparrow, \downarrow, \hat{1}, \hat{0} \right\} \quad \text{— symmetric,}$$

$$\{ \rightarrow, \leftarrow, \backslash, /, \perp, \top, \dashv, \vdash \} \quad \text{— asymmetric.}$$

Of the sixteen logic functions, four have properties which closely resemble the properties of the corresponding functions of one argument:

$$\perp(x,y) = x \perp y = x$$

$$\top(x,y) = x \top y = \bar{x}$$

$$\dashv(x,y) = x \dashv y = y$$

$$\vdash(x,y) = x \vdash y = \bar{y}$$

The reason is that these functions are actually dependent on one argument only; either x:

$$\perp(x,y) = x$$
$$\top(x,y) = \bar{x} \,,$$

or y:

$$\dashv(x,y) = y$$
$$\vdash(x,y) = \bar{y} \,.$$

The connectives $\perp$, $\top$ and $\dashv$, $\vdash$ are converses of each other:

$$x \perp y = y \dashv x$$
$$x \top y = y \vdash x \,.$$

They behave as unary constants with respect to one argument and as filters with respect to the other, which is why they are referred to as filter and antifilter and their converses respectively.

The function $(x \neq y)$ we have chosen to describe as inequivalence: the value of the function is 1 when the values of the two arguments are different, and it is 0 when their values are the same. On the other hand, exactly the same requirements are assigned to exclusive disjunction, XOR: the latter yields 1 when either one or the other argument, but not both, equals 1. Consequently, we are able to give different interpretations to the same logic function.
Function $(x \setminus y)$ is called inhibition because of the role it plays in engineering applications. The value of the function is equal to the value of the first argument when the value of the second is 0. When the second argument is 1, inhibition yields the value 0 independently of the value obtained by the first argument. For converse inhibition (x / y) these relationships are reversed.

Finally, we shall consider the functions NAND $\downarrow$ and NOR $\uparrow$ which play a unique role in conventional logic. Though the connectives corresponding to these functions can easily be found in ordinary language, their discovery is due to the formal development of symbolic logic and is not the result of the formalization of general logical discourse, as was the case with the connectives AND and OR.
Function NAND, an abbreviation for Not-AND, corresponds in ordinary language to the connective "not both ..."; it yields 0 if and only if both arguments have the value 1. Function NOR, an abbreviation for Not-OR, corresponds to the connective "neither ... nor ..."; it yields 1 if and only if both arguments have the value 0. Sometimes NAND and NOR are denoted by the compound symbols which directly reflect their meaning: NAND as $\overline{\wedge}$, and NOR as $\overline{\vee}$. The

pair ↓ and ↑ is complementary to the pair ∧ and ∨, but there is a great difference between the functions NAND and NOR and AND and OR, which we will consider in greater detail shortly.

Basis sets

Given a set of functions one is usually interested in the possibility of expressing some functions via the others and determining the minimal irreducible set of functions — the basis set. The Boolean functions are no exception and the question naturally arises: what are the basis sets of functions which generate the complete set of logic functions?
The question of the minimal functionally complete set of logic functions is of fundamental significance in logic, both theoretically and with regard to applications.
The set of Boolean functions is redundant, and its reduction leads to interesting outcomes. In order to diminish redundancy we may begin by noticing that half of the set can be obtained from the other half by negation. Indeed,

$$\langle 0 \rangle = \overline{\langle 1 \rangle}$$
$$x \neq y = \overline{x \leftrightarrow y}$$
$$x \setminus y = \overline{x \rightarrow y}$$
$$x \,/\, y = \overline{x \leftarrow y}$$
$$x \downarrow y = \overline{x \wedge y}$$
$$x \uparrow y = \overline{x \vee y}$$
$$x \perp y = \overline{x \top y}$$
$$x \dashv y = \overline{x \vdash y}$$

Hence the operations <0>, ≠, \ , /, ↓, ↑, ⊥,⊣ can be excluded as redundant. Taking into account that

$$x \perp y = x$$
$$x \dashv y = y$$

we can exclude two more redundant logic operations, thus arriving at the following set of logic functions:

constant	1
negation	$\overline{x}$
conjunction	$x \wedge y$
disjunction	$x \vee y$
implication	$x \rightarrow y$
converse implication	$x \leftarrow y$
equivalence	$x \leftrightarrow y$

Noting that converse implication is expressible by implication and negation,

$$x \leftarrow y = \overline{x} \rightarrow \overline{y}$$

we get a set of six logic functions, commonly used in conventional logic as the standard basis set. The standard set $\{1, \urcorner, \wedge, \vee, \rightarrow, \leftrightarrow\}$ is sufficient to express any logic function. However, further reductions are possible. Excluding the constant 1, implication and equivalence by equalities:

$$1 = x \vee \overline{x}$$
$$x \rightarrow y = \overline{x} \vee y$$
$$x \leftrightarrow y = (\overline{x} \vee y) \wedge (x \vee \overline{y})$$

we arrive at a set of elementary logic functions:

$$\{\overline{x}, x \wedge y, x \vee y\} ,$$

which contains only three connectives; NOT, AND and OR:

$$\{\urcorner, \wedge, \vee\} .$$

This set is the major basis set most frequently used in applications.

Minor basis sets

The major basis set allows the reduction of the complete set of logic functions to just three logic functions. Nevertheless, it has been found that further reduction of the major set is possible. Several functionally complete minor basis sets, containing only two logic functions, can be derived with the help of the De Morgan relations:

$$x \vee y = \overline{\bar{x} \wedge \bar{y}}$$

$$x \wedge y = \overline{\bar{x} \vee \bar{y}}$$

and the relation between implication $\rightarrow$ and conjunction $\wedge$ and disjunction $\vee$:

$$x \vee y = \bar{x} \rightarrow y$$

$$x \wedge y = \overline{x \rightarrow \bar{y}}$$

In view of the given equalities any pair of logic functions:

$$\{\neg, \wedge\}, \{\neg, \vee\}, \{\neg, \rightarrow\}$$

represents a functionally complete minor basis set.

<u>Minimal basis sets</u>

The minor basis sets can be further reduced to the minimal basis sets, which implies that connectives exist that are functionally complete in themselves. Of the 20 Boolean connectives two have been found to possess this interesting property: NAND $\downarrow$ and NOR $\uparrow$.
Since negation and conjunction, negation and disjunction, and negation and implication have already been shown to be functionally complete, to prove that NAND or NOR are functionally complete by themselves is to show that the minor basis sets can be expressed in the unit set of NAND or NOR. We could achieve this aim by converting an expression written in one of minor basis sets into an expression written in NAND or NOR. With the help of the following equalities:

$$\bar{x} = x \downarrow x \qquad \bar{x} = x \uparrow x$$

$$x \wedge y = \overline{x \downarrow y} \qquad x \vee y = \overline{x \uparrow y}$$

$$x \vee y = \bar{x} \downarrow \bar{y} \qquad x \wedge y = \bar{x} \uparrow \bar{y}$$

the minor basis set $\{\neg, \wedge\}$ will thus be reduced to the minimal basis set $\{\downarrow\}$ or $\{\uparrow\}$:

$$\bar{x} = (x \downarrow x) \qquad \bar{x} = (x \uparrow x)$$

or:

$$(x \wedge y) = (x \downarrow y) \downarrow (x \downarrow y) \qquad (x \wedge y) = (x \uparrow x) \uparrow (y \uparrow y)$$

Similarly, the minor basis set $\{\neg, \vee\}$ is reducible to the minimal basis sets:

$$\bar{x} = (x \downarrow x) \qquad\qquad \bar{x} = (x \uparrow x)$$

or:

$$(x \vee y) = (x \downarrow x) \downarrow (y \downarrow y) \qquad\qquad (x \vee y) = (x \uparrow y) \uparrow (x \uparrow y)$$

The minimal basis sets $\{\downarrow\}$ and $\{\uparrow\}$ allow maximum economy in connectives.

Unfortunately the economy in connectives is gained at the expense of clarity, for the use of a single connective can give rise to rather complicated expressions. For example, a simple logic expression like

$$x \rightarrow x$$

can, in the minimal basis set $\{\downarrow\}$, take the form:

$$((x \downarrow (x \downarrow x)) \downarrow (x \downarrow (x \downarrow x))) \downarrow ((x \downarrow (x \downarrow x)) \downarrow (x \downarrow (x \downarrow x))).$$

Clearly, in applications it is much more convenient to use major or minor basis sets, because loss of clarity is the price which must usually be paid if minimal basis sets are used. Theoretically, however, the possibility of expressing any logic function by a single connective represents an interesting fact.

Boolean algebra

Since different logic expressions can denote the same logic function, the question frequently arises whether two expressions represent the same Boolean function. The equivalence of logic functions can be verified by comparing their truth tables. This method, however, does not allow one to obtain new equalities and becomes very difficult as the number of variables increases.

The fact that the set of functions which consists of NOT, AND and OR can be used to define the complete set of logic functions led to the creation of the Boolean logic algebra. The latter is based on the following system of axioms:

$$\bar{\bar{x}} = x$$

$$x \wedge x = x \qquad\qquad x \vee x = x$$

$$x \wedge \bar{x} = 0 \qquad\qquad x \vee \bar{x} = 1$$

$$x \wedge 1 = x \qquad\qquad x \vee 1 = 1$$

$$
\begin{array}{ll}
x \wedge 0 = 0 & x \vee 0 = x \\
x \wedge y = y \wedge x & x \vee y = y \vee x \\
x \wedge (y \wedge z) = (x \wedge y) \wedge z & x \vee (y \vee z) = (x \vee y) \vee z \\
x \wedge (y \vee z) = (x \wedge y) \vee (x \wedge z) & x \vee (y \wedge z) = (x \vee y) \wedge (x \vee z) \\
\overline{x \wedge y} = \bar{x} \vee \bar{y} & \overline{x \vee y} = \bar{x} \wedge \bar{y}
\end{array}
$$

Boolean algebra is a structure with two binary operations, AND and OR, a unary operation, NOT, and two constant or nullary operations, 0 and 1. The mapping from logic expressions to real algebraic expressions is a consequence of the idempotency postulate:

$$x^2 = x \,.$$

Because the only positive numbers satisfying the idempotency postulate are 1 and 0, where 1 stands for true and 0 for false, the classical contradiction law follows naturally:

$$x^2 - x = 0$$
$$x(1 - x) = 0$$

This gives us an algebraic rule for negation operation, which is expressed simply as subtraction from 1:

$$\neg(x) = 1 - x \,.$$

The logic operations AND and OR are replaced by the ordinary operations of multiplication and addition respectively:

Logic operations	Algebraic equivalent
$\neg x$	$1 - x$
$\wedge$	$\times$
$\vee$	$+$

Employing these coding procedures, any logic expression can be mapped into a corresponding real algebraic expression. It should be emphasized that such mapping is straightforward only for logic expressions which are formulated in the major basis set or minor basis sets. All connectives, except $\rceil$, $\wedge$, $\vee$, are exempt from the direct coding procedure; and an appropriate transformation of the expression is necessary prior to algebraic mapping, should it contain other logic connectives.

One of the most serious defects of conventional logic algebra is the artificial rule for logic addition:

$$1 + 1 = 1$$

Though useful in computer applications, this physically unnatural convention broadens, rather than narrows, the gap between logic and the exact sciences.

Another point to be mentioned concerns equivalent logic expressions. In general, two expressions that are logically equivalent will not be mapped to the same algebraic expression or even equivalent real algebraic expressions. In other words, Boolean mapping is not a genuine mapping to real functions. Nevertheless, applying the system of equalities of the Boolean logic algebra, we can now deal with various problems on a purely formal basis, and for many of them effective solutions can be formulated.

Like the arithmetic operations of ordinary algebra, logic operations in Boolean algebra have different priorities, which are reflected in the order of their execution in expressions. First, NOT is executed, followed by AND, and only then is OR performed. In cases where this order is changed, brackets are used as in algebraic expressions.

The duality principle

An important principle, which reduces by half the effort involved in the derivation of logic equalities, is the duality principle. The binary logic connectives form the following dual relationships:

$$\begin{array}{ccc}
\leftrightarrow & - & \neq \\
\wedge & - & \vee \\
\uparrow & - & \downarrow \\
\rightarrow & - & / \\
\leftarrow & - & \backslash \\
\perp & - & \perp \\
\mathrm{T} & - & \mathrm{T} \\
\dashv & - & \dashv \\
\vdash & - & \vdash \\
\langle 0 \rangle & - & \langle 1 \rangle
\end{array}$$

To form a dual formula X* from a formula X, the symbols for connectives and variables must be replaced by their duals. Then, according to the duality principle, a formula X and its dual X* are related as follows:

$$\overline{X} = X^* ,$$

$$X = \overline{X^*} .$$

If X = Y and X* is dual to X, and Y* to Y, then X* = Y*.
Examples of dual transformation can be found in the De Morgan laws:

$$\overline{x \wedge y} = \bar{x} \vee \bar{y} ; \qquad \overline{x \vee y} = \bar{x} \wedge \bar{y} ,$$

and in the pair of distributive laws:

$$x \wedge (y \vee z) = (x \wedge y) \vee (x \wedge z),$$

$$x \vee (y \wedge z) = (x \vee y) \wedge (x \vee z).$$

The contradiction law

$$x \wedge \bar{x} = 0$$

is dual to the law of the excluded middle:

$$\bar{x} \vee x = 1 .$$

The duality principle enables us to construct new equalities without referring to the truth-tables. And, given a basis set of equalities, any equality of the system can be deduced from it by a series of equivalent transformations. The equivalent transformations change the formula but not the logic function realized by this formula.

Tautologies and contradictions

Logic expressions which are true under any interpretation of their constituent variables are called tautologies. Logic expressions which are false under any interpretation of their constituent variables are called contradictions. In the resulting truth-table for tautology there are no

other entries but 1, whereas contradiction yields only 0. Clearly, tautology and contradiction can be related by negation.
For example, from the contradictions $\bar{x} \wedge x$, $x \leftrightarrow \bar{x}$, and $x \setminus x$ one derives the tautologies:

$$\overline{\bar{x} \wedge x} = x \vee \bar{x}$$

$$\overline{x \leftrightarrow \bar{x}} = x \leftrightarrow x$$

$$\overline{x \setminus x} = x \to x$$

Thus, an expression is a tautology if and only if its negation is a contradiction. If X and $X \to Y$ are tautologies, then Y is a tautology. If $X \wedge Y$ is a tautology, then X and Y are tautologies. If X is a tautology, then $X \vee Y$ and $Y \vee X$ are tautologies. If $X \to Y$ is a tautology, then $X^* \to Y^*$ is a tautology. If $X \leftrightarrow Y$ is a tautology, then $X^* \leftrightarrow Y^*$ is a tautology.
An expression which contains only a connective $\leftrightarrow$ is a tautology if and only if each variable enters an expression an even number of times.

Normal forms

The evaluation of logic formulas is further simplified by their reduction to a standard normal form. The fundamental theorem of Boolean algebra states that any logic formula can be expressed in the disjunctive normal form or conjunctive normal form, defined as follows.
An expression is said to be in disjunctive normal form if it represents a disjunction of one or more disjuncts, each of which is a continued conjunction of atomic variables or their negations:

$$\bigvee_{i=1}^{m} \left(\bigwedge_{j=1}^{n} x_{ij} \right)$$

Any expression which is not a contradiction is logically equivalent to some disjunctive normal form. An expression is said to be in conjunctive normal form if it represents a conjunction of one or more conjuncts, each of which is a continued disjunction of atomic variables or their negations:

$$\bigwedge_{i=1}^{m} \left(\bigvee_{j=1}^{n} x_{ij} \right)$$

Any expression which is not a tautology is logically equivalent to some conjunctive normal form.

A single variable or its negation can be considered as a degenerate disjunction or conjunction. Every disjunctive normal form is dual to some conjunctive normal form. To obtain a dual form we replace connectives by their duals and interchange 0 and 1.

$$\bigvee_{i=1}^{m} (\bigwedge_{j=1}^{n} x_{ij}) = \overline{\bigwedge_{i=1}^{m} (\bigvee_{j=1}^{n} \bar{x}_{ij})}$$

Any formula is logically equivalent to some disjunctive or conjunctive normal form, and can be reduced to such a form, either directly from its truth-table or by using the method of equivalent transformations. The interpretation under which the full disjunctive normal form becomes true can be read from it directly. Another advantage lies in the ease with which validity is recognizable.

Disjunctive and conjunctive normal forms are used to solve the problems of finding all the hypotheses and consequents of a given formula. Formula X is a hypothesis for formula Y if

$$X \rightarrow Y = 1,$$

and formula Z is a consequent of formula Y if

$$Y \rightarrow Z = 1.$$

A hypothesis X of formula Y is simple if it is a conjunction of atomic variables or their negations and ceases to be a hypothesis of formula Y after deletion of any of the conjuncts. Similarly, a consequent Z of formula Y is simple if it is a disjunction of atomic variables or their negations and ceases to be a consequent after deletion of any of the disjuncts.

The solution to the problem of finding hypotheses and consequents requires the construction of all the simple hypotheses and consequents for a given formula, and then obtaining all other hypotheses and consequents with the help of equivalent transformations.

A disjunct of a disjunctive normal form is a hypothesis of this disjunctive normal form; a conjunct of a conjunctive normal form is a consequent of this disjunctive normal form.

A full disjunctive normal form has no other hypothesis but the disjunctions of some of its disjuncts. A full conjunctive normal form has no other consequents but the conjunction of some of its conjuncts.

A reduced disjunctive normal form is a disjunction of its simple hypotheses; a reduced conjunctive normal form is a conjunction of its simple consequents.

A reduced disjunctive normal form has important applications for the minimization of logic functions. The minimization of logic functions requires the construction of a disjunctive normal form which has a minimal number of conjuncts in its disjuncts, and thus minimal complexity.

1.1.2 Multivalued logic

In multivalued logics the class of logic functions is expanded to incorporate those functions whose arguments, as well as the functions themselves, are defined in more than a binary domain. The primary objective of multivalued logic is to devise a set of truth-values that encodes intermediate levels of validity between true or valid and false or invalid. Clearly, the introduction of multivalued logic systems is possible only at the expense of rejecting the universal validity of the classical law of the excluded middle:

$$x \vee \bar{x} = 1,$$

which gives rise to ternary or three-valued logic; the excluded 4th, which gives rise to the tetrary or four-valued logic, or, generally, the excluded kth which allows one to expand (k–1)-valued logic into a k-valued logic.

In expanding the alphabet of truth-values, multivalued logic represents a generalization of Boolean two-valued logic. The multivalued logic funtions are defined on the set

$$E_k = \{0, 1, \dots, k-1\} \; ; \quad k = 2, 3, \dots$$

which for $k = 2$ becomes the binary set $E_2 = \{0,1\}$ of two-valued logic. The number of k-valued functions of n arguments $x_1, x_2, \dots x_n$ is k^{k^n}. For $k \geq 3$, the difficulties of truth-table presentation and effective enumeration of all possible functions, in comparison with binary logic, increase rapidly. In three-valued logic there are $3^9 = 19683$ possible truth functions with two arguments, a significant difference from two-valued logic, where there are only 16 different functions with two arguments.

It is quite understandable that multivalued logic systems have been attracting growing attention for their wealth of logic structures. The elementary functions of binary logic have several analogues in multivalued systems. Consider some generalizations of negation:

$$\bar{x} = x + 1 \pmod{k} \quad - \quad \text{cyclic negation,}$$

$$\sim x = k - 1 - x \quad - \quad \text{diametrical negation.}$$

Cyclic negation is a generalization of negation involving a cyclic shift of values. Another aspect of negation is generalized by diametrical negation, which is general multivalued complementation.

There are also several generalizations for the logic functions of two arguments. The most commonly used generalizations of conjunction and disjunction are the following:

Conjunction	Disjunction
a) min (x,y)	a) max (x,y)
b) $x \cdot y \pmod k$	b) $x + y \pmod k$.

The multivalued functions in many respects resemble the functions of binary logic and some results of the latter remain applicable in multivalued logic. However, there are important novel properties which distinguish a logic for $k \geq 3$ from classical binary logic. Consider (for $k \geq 3$):

$$\sim(\sim x) = x \quad \text{but} \quad \bar{\bar{x}} \neq x$$

$$\sim \min(x,y) = \max(\sim x, \sim y) \quad \text{but} \quad \overline{\min(x,y)} \neq \max(\bar{x}, \bar{y}) \ .$$

An important class of multivalued functions is represented by the Post functions:

$$J_i(x) = \begin{cases} k-1 & \text{for} \quad x = i \\ 0 & \text{for} \quad x \neq i \end{cases}$$

$$i = 0, k-1 \ .$$

The functions $J_i(x)$ allow reduction of multivalued functions to normal forms, the analogue of fundamental reduction in binary Boolean logic to disjunctive normal forms.

The three-valued approach introduces the intermediary truth-value in addition to true and false. With the third truth-value added to the set of truth-values, the truth-tables for $\wedge$ and $\vee$ take the following form, where $\frac{1}{2}$ denotes the indeterminate truth-value:

x \ y	0	$\frac{1}{2}$	1
0	0	0	0
$\frac{1}{2}$	0	$\frac{1}{2}$	$\frac{1}{2}$
1	0	$\frac{1}{2}$	1

Table 1.1 Three-valued conjunction

x \ y	0	$\frac{1}{2}$	1
0	0	$\frac{1}{2}$	1
$\frac{1}{2}$	$\frac{1}{2}$	$\frac{1}{2}$	1
1	1	1	1

Table 1.2 Three-valued disjunction

An attractive feature of this approach is that the truth-tables of three-valued conjunction and disjunction, with deletion of the indeterminate truth-value $\frac{1}{2}$, are reduced to the corresponding classical truth-tables.

The question of the relation between binary and multivalued logics is not as simple as it may appear. Multivalued logic denies the universal validity of the contradiction law of classical logic. This revision of classical logic is adopted in order to accommodate the logic operations which display properties that do not conform to the classical binary rules of inference. Multivalued logic retains the validity of logic theory at the expense of revising the roles of the connectives, which are altered so that compliance with the contradiction law is no longer the universal requirement for logic relations. However, an assumption that multivalued logic is more general than binary logic and contains the latter as a degenerate case runs counter to our understanding of the role of classical binary logic, the logic we actually use to describe multivalued logics. Since multivalued logic is rooted in binary reasoning, which makes extensive use of classical logic, it seems a contradiction in terms that this theory can give rise to a revision of logic which denies classical logic itself.

1.2 Probability logic

Deterministic logic covers aspects of logic that are of great theoretical and applied significance. However, there are logical truths that lie wholly beyond the grasp of deterministic logics. Approximate indeterministic forms of reasoning are an important aspect of the operations of the mind. A broader analysis of logic operations is achieved in modal logic, which deals with the concepts of necessity and possibility. This requires the introduction of modal connectives, $\Box$x, read as 'necessarily x'; and $\Diamond$ x, read as 'possibly x'. Since necessarily $\bar{x}$, $\Box\,\bar{x}$, makes precisely the same assertion as impossible x, $\neg\Diamond$ x, whereas not necessarily x, $\neg\Box$x, is equivalent to possibly $\bar{x}$, $\Diamond\,\bar{x}$, we deduce the following fundamental relationships between the modal connectives:

$$\Box = \neg\Diamond\neg$$

$$\Diamond = \neg\Box\neg$$

The connectives $\Box$ and $\Diamond$ are interchangeable and the expressive power of modal logic is in no way impaired by the omission of either of them.

The computational analysis of modal logic can be enhanced with the support of probability theory. Indeed, the abstract concept of possibility is quite naturally translated into the rigorous concept of probability, which in turn gives rise to probability logic. Probability logic is a straightforward application of probability algebra to logic expressions, interpreted as indeterministic events.

An expression x is characterized by probability of realization P(x), that is, a strictly positive, normalized and additive function taking its values in the unit interval:

$$0 \leq P(x) \leq 1.$$

A true expression is then an expression with probability 1,

$$P(x) = 1.$$

A false expression is characterized by a probability equal to 0,

$$P(x) = 0.$$

Naturally, certain realization Ω is a tautology,

$$P(\Omega) \equiv 1\,,$$

and impossible realization is a contradiction,

$$P(\overline{\Omega}) = P(\varnothing) \equiv 0.$$

Since we associated truth and falsity with the probability of realization of a logic expression, realizations with probabilities 1 or 0 are particular cases of realization with probability

$$0 \leq P(x) \leq 1,$$

and thus classical logic can be considered to represent a special case of probability logic. Whenever logical truth or falsity cannot be established with certainty, the probability function is confined to the unit interval excluding its boundary:

$$0 < P(x) < 1.$$

By assigning to atomic expressions some probability of realization, we assume that a compound expression can also be characterized by a probability of realization, which depends on the probabilities of realization of the atomic expressions. The investigation of these dependencies is the primary target of probability logic.

1.2.1 Negation

The algebras behind deterministic binary negation and modal negation are quite similar, because both operations represent the algebraic complement.
Consider the complete set Ω comprising just two disjoint expressions, so that the realization of one of them precludes the occurrence of the other. Such expressions are defined as opposite or complementary. If one is denoted by x, then $\overline{x}$ denotes the other, and the sum of their probabilities is a tautology:

$$P(x) + P(\overline{x}) = P(\Omega) \equiv 1$$

From this it follows that:

$$P(x) + P(\bar{x}) = 1$$

$$P(\bar{x}) = 1 - P(x) = \overline{P(x)}$$

This establishes the negation operation in probability logic

$$\overline{P(x)} = P(\bar{x}).$$

For modal negation the complement of probability equals the probability of complementary realization. When the probability of realization of an expression is known, the probability of the opposite or complementary realization is found by modal negation (complementation). In particular, if both expressions have the same probability of realization it is equal to $\frac{1}{2}$.

The transition from modal negation to Boolean negation occurs in limiting cases. Whenever the probabilities obtain only the limit values 1 or 0, then the events are absolutely certain or impossible, and modal negation turns into the negation of classical logic.

1.2.2 Conjunction

Whereas in Boolean logic the truth function of a compound expression does not depend on the content of the atomic expressions, in probability logic the content of the atomic expressions can change essentially the probability of a compound. This brings into focus the question of the independence of logic expressions, which is of major importance in probability logic. While Boolean logic is interested only in the truth status of expressions, it is a novel and crucial task of probability logic to examine the conditions under which modal expressions can be considered to be dependent or independent. The concept of independence is particular to probability logic, and it is this concept which really marks off probability logic as a distinct branch of logic theory.
Modal expressions are dependent when they are not free to occur in an arbitrary combination or sequence, that is, some compound expression depending upon them is precluded from occurring. Here the range of what is possible is less than the range of what is conceivable.
The subject of logic independence can be best understood through an examination of modal operations. In calculating the probability of modal conjunction we must take into account whether the logic expressions are independent or not. Depending on the answer to this question the value of modal conjunction will be different.

For independent expressions the truth or falsity of one is in no way connected with the truth or falsity of the other, and modal conjunction takes the following form:

$$P(x \wedge y) = P(x)\,P(y)\,.$$

Thus, provided the expressions are independent, the probability of a product is the product of the probabilities. But if the independence condition is not satisfied, conditional probability is required for the evaluation of modal conjunction:

$$P(x \wedge y) = P(x)\,P(y \mid x)\,,$$
$$P(x \wedge y) = P(y)\,P(x \mid y)\,.$$

In the special case that

$$P(y \mid x) = P(y)$$
$$P(x \mid y) = P(x)$$

both conjunctions give exactly the same result. Consequently, the expressions are independent if and only if the conditional probabilities equal the absolute probabilities. It is in this case that y is independent of x and vice versa, and the question of the truth of x is irrelevant for determining the probability of y and vice versa. This establishes the necessary and sufficient condition for independence.

A transition from probability conjunction to Boolean conjunction is achieved by employing the limiting probabilities 1 or 0, so that the latter is a particular case of probability conjunction. Modal logic is not only more general than Boolean logic, it also calls for an investigation of the interdependencies between logic expressions. It is very important to recognize that in modal logic the probabilities may not be arbitrary values between 0 and 1 but are frequently subject to restrictions; not all values are consistent. Whenever

$$P(x),\, P(y) \neq 1, 0\,,$$

the value of modal conjunction is less than the probabilities of each of the constituents:

$$P(x \wedge y) < P(x),\, P(y)\,.$$

When $P(x) = 1$ and $P(y) \neq 1, 0$, the probability of realization of a compound is

$$P(x \wedge y) \neq 1, 0 .$$

When

$$P(x) = p ,$$

then

$$P(x \wedge y) = q$$

would not be a possible set of data unless $q \leq p$.

Modal conjunction can be extended by induction to any finite number of expressions:

$$P(x_1 \wedge x_2 \wedge \ldots \wedge x_n) = P(x_1)\, P(x_2) \ldots P(x_n) .$$

1.2.3 Disjunction

Whereas for Boolean disjunction it does not matter whether the constituent expressions are connected by virtue of their content, modal disjunction does depend on the content of the constituent expressions. The value of modal disjunction can be found by the law of addition of probabilities:

$$P(x \vee y) = P(x) + P(y) .$$

This relation holds only if the constituent expressions are disjoint, otherwise modal disjunction must be modified to account for the inclusive part of the expressions:

$$P(x \vee y) = P(x) + P(y) - P(x \wedge y) .$$

A conjunction probability $P(x \wedge y)$ occurs in the formulation of modal disjunction as the subtraction term which eliminates the overlap wherever present. When x and y are mutually exclusive, then $P(x \wedge y) = 0$ and the general rule agrees with the special case.

When we derived modal negation above, we in fact employed modal disjunction.
Since $P(x \vee \overline{x})$ is a tautology

$$P(x \vee \overline{x}) = 1$$

and x and $\overline{x}$ are mutually exclusive, the addition law applies

$$P(x \vee \overline{x}) = P(x) + P(\overline{x}) ,$$

and

$$P(x) + P(\overline{x}) = 1 ,$$

which establishes the negation rule in probability logic:

$$\overline{P(x)} = P(\overline{x}) .$$

An important distinction between modal disjunction and modal conjunction is that the latter can never be defined in strictly binary logic terms, unless all its constituents are also defined as certain or impossible events.

In contrast, even when the probabilities of realization of constituents of modal disjunction are less than 1, the disjunction can still yield 1, and therefore will be true in the deterministic logic sense. Remarkably enough, this takes place when the constituents are not defined in terms of true or false in a deterministic sense.

If modal disjunction is applied to the expressions comprising the complete set Ω, the resulting probability value is 1, and vice versa, if the modal disjunction is a tautology, the expressions form the complete set Ω:

$$P(x_1 \vee x_2 \vee \ldots \vee x_n) = P(\Omega) .$$

Due to the additive property of probabilities, whenever the expressions are disjoint, the probability of disjunction is a sum of the probabilities:

$$P(x_1 \vee x_2 \vee \ldots \vee x_n) = P(x_1) + \ldots + P(x_n) .$$

For a complete set modal disjunction is a tautology:

$$P(x_1) + P(x_2) + \ldots + P(x_n) \equiv 1 .$$

1.2.4 Conditional

The construction of the modal conjunction, disjunction and negation functions was a straightforward application of the basic probability laws of multiplication, addition and complementation respectively. The form of the modal conditional can be obtained from the basic laws as well.

Because modal conjunction generally includes conditional probabilities, the general form of the multiplication law

$$P(x \wedge y) = P(x)\, P(y \mid x)$$
$$P(x \wedge y) = P(y)\, P(x \mid y)$$

can be taken as a definition of the modal conditional

$$P(x \text{ if } y) = P(x \mid y)\,.$$

Explicitly we have

$$P(y \mid x) = \frac{P(x \wedge y)}{P(x)} \qquad P(x) \neq 0$$

$$P(x \mid y) = \frac{P(x \wedge y)}{P(y)} \qquad P(y) \neq 0\,.$$

It is easy to establish that

$$P(y \mid x) \geq 0$$
$$P(\Omega \mid x) = 1$$
$$P(x \mid x) = 1$$
$$P(x)\, P(y \mid x) = P(y)\, P(x \mid y)\,.$$

From the last equality we obtain the important form of the modal conditional, which allows us to reassess the validity of the premisses employed in probabilistic reasoning:

$$P(x \mid y) = \frac{P(x)\, P(y \mid x)}{P(y)} = $$
$$= \frac{P(x)\, P(y \mid x)}{P(x)\, P(y \mid x) + P(\overline{x})\, P(y \mid \overline{x})}\,,$$

where P(y) is expressed in accordance with the theorem of total probability:

$$P(y) = P(x)\,P(y \mid x) + P(\overline{x})\,P(y \mid \overline{x})\ .$$

Whenever the expressions are independent,

$$P(y \mid x) = P(y \mid \overline{x})\ ,$$

which implies $P(x \mid y) = P(x)$.

The modal conditional plays an important role in probability logic. However, because conditional probability is not defined for the limiting case of a false antecedent, $P(y) = 0$, it cannot in the general case serve as a model of modal implication. Although the modal conditional accounts for some properties of modal implication, it does not cover them in their totality. A principal difference between conditional probability and the probability of the conditional has to be duly recognized.

CHAPTER 2

MATRIX LOGIC

Logic mirrors in its structure the fundamental properties of the mind. Therefore the mathematization of logic should not be considered a purpose in itself. Because logic is the embodiment of high-level intelligence, the deciphering of the laws and the code of the mind is greatly dependent on our ability to derive logic theory from fundamental principles. A challenging goal is to incorporate the phenomena of intelligence into the general system of the covariant laws of nature.

The matrix operator formulation of logic, developed further, allows us to reduce the study of the laws of logic to an investigation of the properties of logic space, much in the same way as the study of Hilbert and Minkowski spaces serves as a key to the study of quantum and relativity laws.

2.1 Logic space

The theory of vector spaces is of fundamental significance in much of modern mathematics. Though conventional logic has succeeded in providing a formalization for certain aspects of deductive reasoning, there remains a vast unexplored area of logic phenomena whose understanding is closely linked with its mathematization. This chapter introduces a new concept of the logic vector spaces that not only enables logic to function fully as an instrument of computation but also brings a new and profound understanding of logic phenomena.

We consider two adjoint spaces of logic vectors, the row or bra-vectors whose components are ordered horizontally:

$$< p\,| = (\bar{p}, p)$$

and the column or ket-vectors whose components are ordered vertically:

$$|\, q > \;=\; \begin{pmatrix} \bar{q} \\ q \end{pmatrix}$$

The key property of logic vectors is the logic normalization condition:

$$\bar{p} + p = 1 .$$

This interdependence of the components of logic vectors is particular to logic space theory, and it is this which really distinguishes logic space as a distinct kind of vector space.
The logic vectors will be used to characterize logic states with the components $\bar{p}$, $\bar{q}$ representing falsum and the components p, q representing verum. The possible values that the components can take are confined to the unit interval

$$0 \leq \bar{p}, \bar{q}, p, q \leq 1 .$$

The bra and ket logic vectors are exchangeable by transposition:

$$\widetilde{|p>} = < p | .$$

Whatever the vector $|p>$ stands for, the same information is held within $<p|$; both are equally informative and neither is to be considered more important or fundamental than the other. The results derived for a ket space are automatically applicable to a bra space, a fact we shall not mention unless the need arises.

The complete logic space can be generated from basis vectors. A basis ν for a logic space Ω is an ordered set of unit vectors

$$\nu = \{|e_1> , |e_2>\} = \left\{ \begin{pmatrix} 1 \\ 0 \end{pmatrix}, \begin{pmatrix} 0 \\ 1 \end{pmatrix} \right\},$$

which span Ω and are linearly independent. The basis vectors may be anything from the unit vectors to the complete set of eigenvectors of an operator. Once the basis is stipulated, we can specify any vector by stating the ordered set of components

$$|p> = \bar{p}|e_1> + p|e_2> = \bar{p}\begin{pmatrix} 1 \\ 0 \end{pmatrix} + p\begin{pmatrix} 0 \\ 1 \end{pmatrix},$$

where $\{\bar{p}, p\}$ is the representation of a vector $|p>$ in the $\{|e_1>, |e_2>\}$ basis system. The vectors can be described equally well in terms of a complementary basis system $\bar{\nu} = \left\{ |\bar{e}_1> , |\bar{e}_2> \right\}$ which is orthogonal to the first:

$$< v \mid \bar{v} > = < e_1 \mid \bar{e_1} > + < e_2 \mid \bar{e_2} > = 0 \, .$$

Thus the complement operation recombines basis vectors and exchanges the falsum and verum components of the logic vector:

$$\overline{\begin{pmatrix} \bar{p} \\ p \end{pmatrix}} = \begin{pmatrix} p \\ \bar{p} \end{pmatrix}.$$

A vector space may not have a finite basis. But if it has a basis with n elements, then any other basis also has n elements. The number n characterizes the dimension of the vector space. A maximal subset of linearly independent vectors is automatically a basis. It follows that if logic space Ω has dimension n then any n vectors which span Ω form a basis.

Closure relation

Special importance attaches to representations of logic vectors in complete orthonormal basis sets, which display orthonormality:

$$<e_i \mid e_j> = \delta_{ij} \, ,$$

and closure:

$$\sum_{i=1}^{2} \mid e_i > < e_i \mid \; = I \, ,$$

where I denotes the unit matrix.

If $\{\mid e_i>\}$ is a complete orthonormal basis set, then we can express vector $\mid p>$ in terms of its components as follows:

$$\mid p > = \sum_{i=1}^{2} p_i \mid e_i > ,$$

where p_i is the orthogonal projection of $\mid p>$ along the basis vector $\mid e_i>$:

$$p_i = <e_i \mid p> .$$

Consequently, the vector $| p>$ can be written

$$| p> = \sum_{i=1}^{2} | e_i >< e_i | p> .$$

By taking a decomposition of $| p>$ relative to some given complete orthonormal basis and summing all the projections of $| p>$ we again get the vector $| p>$, which leads immediately to the closure relation:

$$I = \sum_{i=1}^{2} | e_i >< e_i | .$$

The closure relation applies if and only if the basis set is complete.

We shall make particular use of basis logic vectors, which are both normalized and orthogonal. In view of the interpretation given to the components of logic vectors, the basis vectors are vector representations of the true and false logic values. Consequently, we consider true and false as orthogonal quantities in a direct mathematical sense. The inner product offers a way of defining the orthogonality of vectors that is both easy and general:

$$<x | y> = \delta_{xy} ,$$

where δ_{xy} is the Kronecker delta defined as follows:

$$\delta_{xy} = \begin{cases} 1 & \text{for } x = y \\ 0 & \text{for } x \neq y \end{cases}$$

Because in logic space the true and the false logic vectors coincide with the basis vectors, an extension of a logic vector as a vector sum of true and false is always an extension in the true and false coordinate system.

The orthogonal projections of a logic vector $| p> = (\overline{p}, p)$ cut the corresponding verum and falsum unit intervals into p and $\overline{p}$, and $\overline{p}$ and p sections respectively. With rotation of a vector, or equivalently with rotation of logic space, the end of the vector runs along the sides of the fundamental normalization square as shown in Fig.1:

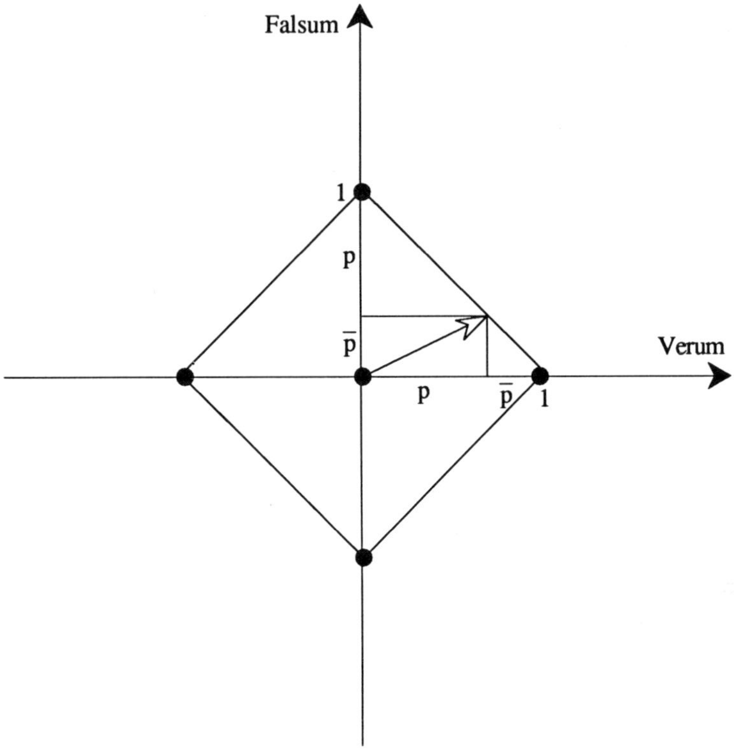

Fig. 1 Logic coordinate system

In this chapter we study only positive-valued logic Ω^+, which must be associated with the upper right section of the Cartesian coordinates, where logic normalization holds:

$$\bar{p} + p = 1 .$$

A more general form of logic normalization by modulo,

$$|\bar{p}| + |p| = 1 ,$$

is considered in the next chapter.

The logic normalization condition implies non-linearity for a transformation of a logic vector. Generally a logic state cannot be translated linearly into another logic state without a violation of logic normalization. The logic normalization condition thus imposes a severe restriction on the multiplication of vectors by scalars.

Another fundamental feature of logic spaces, the operation of addition for logic vectors, is not generally defined, because such an operation could take us outside the logic space of correctly normalized vectors, where logical interpretation is not available. This effectively forbids the addition of the basis vectors of the same truth status:

$$\begin{pmatrix}1\\0\end{pmatrix} + \begin{pmatrix}1\\0\end{pmatrix} = \begin{pmatrix}2\\0\end{pmatrix} \quad \text{— forbidden}$$

$$\begin{pmatrix}0\\1\end{pmatrix} + \begin{pmatrix}0\\1\end{pmatrix} = \begin{pmatrix}0\\2\end{pmatrix} \quad \text{— forbidden}$$

Logic operators

Our objective is to describe the processes of logic evaluations in terms of logic space transformations. A crucial step in achieving the stated objective is the construction of logic space operators.

A logic operator describes a mapping conveniently expressed by the equation

$$| y> = L | x> ,$$

where L is an operator, $| x>$ and $| y>$ are logic vectors, not necessarily normalized. Consequently we may present the meaning of L by listing each $| x>$ of Ω, and then the corresponding $| y> = L | x>$. But we can much more efficiently define the same operator L by recording what it does to the basis vectors spanning the logic space Ω. With the aid of the closure relation we obtain

$$L | e_j > = \sum_{i=1}^{2} | e_i > < e_i | (L | e_j >) .$$

Because no special importance is ascribed either to bra or to ket vector space, we must assume that $<e_i | L | e_j>$ also has a meaning. The operator L was defined through the way it affects the ket basis vectors. But we may equally well consider that the operator acts on bra basis vectors:

$$(<e_i | L) | e_j> = <e_i | (L | e_j>) = <e_i | L | e_j> .$$

Consequently, the priority brackets can be freely omitted and the operator can operate either to the left or to the right, as the need arises.

The initial set of logic operators we obtain directly from the truth-tables for logic connectives. Instead of viewing the truth-tables as a convenient illustration introduced simply for the purpose of friendly usage, we interpret them as mathematical objects, namely as (2×2) matrix operators defined as follows:

Name	Symbol	L	$\tilde{L}$	$\overline{L}$
Equivalence	$\leftrightarrow$	$\begin{pmatrix}1 & 0\\0 & 1\end{pmatrix}$	$\begin{pmatrix}1 & 0\\0 & 1\end{pmatrix}$	$\begin{pmatrix}1 & 0\\0 & 1\end{pmatrix}$
Negation	$\urcorner$	$\begin{pmatrix}0 & 1\\1 & 0\end{pmatrix}$	$\begin{pmatrix}0 & 1\\1 & 0\end{pmatrix}$	$\begin{pmatrix}0 & 1\\1 & 0\end{pmatrix}$
Conjunction	$\wedge$	$\begin{pmatrix}0 & 0\\0 & 1\end{pmatrix}$	$\begin{pmatrix}0 & 0\\0 & 1\end{pmatrix}$	$\begin{pmatrix}1 & 0\\0 & 0\end{pmatrix}$
Disjunction	$\vee$	$\begin{pmatrix}0 & 1\\1 & 1\end{pmatrix}$	$\begin{pmatrix}0 & 1\\1 & 1\end{pmatrix}$	$\begin{pmatrix}1 & 1\\1 & 0\end{pmatrix}$
Implication	$\rightarrow$	$\begin{pmatrix}1 & 1\\0 & 1\end{pmatrix}$	$\begin{pmatrix}1 & 0\\1 & 1\end{pmatrix}$	$\begin{pmatrix}1 & 1\\0 & 1\end{pmatrix}$
Converse implication	$\leftarrow$	$\begin{pmatrix}1 & 0\\1 & 1\end{pmatrix}$	$\begin{pmatrix}1 & 1\\0 & 1\end{pmatrix}$	$\begin{pmatrix}1 & 0\\1 & 1\end{pmatrix}$
Inhibition	$\backslash$	$\begin{pmatrix}0 & 0\\1 & 0\end{pmatrix}$	$\begin{pmatrix}0 & 1\\0 & 0\end{pmatrix}$	$\begin{pmatrix}0 & 0\\1 & 0\end{pmatrix}$
Converse inhibition	$/$	$\begin{pmatrix}0 & 1\\0 & 0\end{pmatrix}$	$\begin{pmatrix}0 & 0\\1 & 0\end{pmatrix}$	$\begin{pmatrix}0 & 1\\0 & 0\end{pmatrix}$
Nor	$\uparrow$	$\begin{pmatrix}1 & 0\\0 & 0\end{pmatrix}$	$\begin{pmatrix}1 & 0\\0 & 0\end{pmatrix}$	$\begin{pmatrix}0 & 0\\0 & 1\end{pmatrix}$
Nand	$\downarrow$	$\begin{pmatrix}1 & 1\\1 & 0\end{pmatrix}$	$\begin{pmatrix}1 & 1\\1 & 0\end{pmatrix}$	$\begin{pmatrix}0 & 1\\1 & 1\end{pmatrix}$
Filter	$\perp$	$\begin{pmatrix}0 & 0\\1 & 1\end{pmatrix}$	$\begin{pmatrix}0 & 1\\0 & 1\end{pmatrix}$	$\begin{pmatrix}1 & 0\\1 & 0\end{pmatrix}$
Antifilter	$\top$	$\begin{pmatrix}1 & 1\\0 & 0\end{pmatrix}$	$\begin{pmatrix}1 & 0\\1 & 0\end{pmatrix}$	$\begin{pmatrix}0 & 1\\0 & 1\end{pmatrix}$
Converse filter	$\dashv$	$\begin{pmatrix}0 & 1\\0 & 1\end{pmatrix}$	$\begin{pmatrix}0 & 0\\1 & 1\end{pmatrix}$	$\begin{pmatrix}1 & 1\\0 & 0\end{pmatrix}$
Converse antifilter	$\vdash$	$\begin{pmatrix}1 & 0\\1 & 0\end{pmatrix}$	$\begin{pmatrix}1 & 1\\0 & 0\end{pmatrix}$	$\begin{pmatrix}0 & 0\\1 & 1\end{pmatrix}$
Tautology	$\hat{1}$	$\begin{pmatrix}1 & 1\\1 & 1\end{pmatrix}$	$\begin{pmatrix}1 & 1\\1 & 1\end{pmatrix}$	$\begin{pmatrix}1 & 1\\1 & 1\end{pmatrix}$
Empty operator	$\hat{0}$	$\begin{pmatrix}0 & 0\\0 & 0\end{pmatrix}$	$\begin{pmatrix}0 & 0\\0 & 0\end{pmatrix}$	$\begin{pmatrix}0 & 0\\0 & 0\end{pmatrix}$

In the chart $\widetilde{\mathrm{L}}$ denotes matrix transposition, which is performed against the principal matrix diagonal (i, i), and $\overline{\mathrm{L}}$ denotes contraposition, that is, converse transposition, which is performed against the secondary matrix diagonal (i, n–i+1). Converse transposition is a new matrix operation, introduced by the author, as a result of a reverse application of logic to matrix theory.

The logic operators are defined in the standard basis ν. However, it has to be clearly understood that there is nothing absolute about the chosen structure of logic operators. In another basis system the structure of logic operators could very well be different. With the inversion of logic space that occurs as we pass from the standard basis ν to the complementary basis $\bar{\nu}$, the logic operators are modified as follows:

L(ν)	L($\bar{\nu}$)	
$\leftrightarrow$ $\urcorner$ $\hat{1}$ $\hat{0}$	$\leftrightarrow$ $\urcorner$ $\hat{1}$ $\hat{0}$	L($\bar{\nu}$) = L(ν)
$\wedge$ $\vee$ $\downarrow$ $\uparrow$	$\uparrow$ $\downarrow$ $\vee$ $\wedge$	L($\bar{\nu}$) = $\overline{\mathrm{L}}$(ν)
$\rightarrow$ $\leftarrow$ $\backslash$ $/$	$\leftarrow$ $\rightarrow$ $/$ $\backslash$	L($\bar{\nu}$) = $\widetilde{\mathrm{L}}$(ν)
$\perp$ $\top$ $\dashv$ $\vdash$	$\top$ $\perp$ $\vdash$ $\dashv$	L($\bar{\nu}$) = $\bar{\mathrm{L}}$(ν)

Examination of the operators obtained shows that they are divided into 4 quadruplets of operators, each reacting differently to the complementation of the basis. In the first quadruplet, complementation is an identity operation; in the second, it is contraposition; in the third, transposition; and in the fourth, matrix complement.

The four distinctly different responses of the operators to the rotation of the logic space establish four different subsets of logic operators:

bisymmetric operators $\{I, \rceil, \hat{1}, \hat{0}\}$

$$L(v) = L(\bar{v}),$$

symmetric operators $\{\wedge, \vee, \downarrow, \uparrow\}$

$$L(v) = \overline{\rceil L}(\bar{v}),$$

contrasymmetric operators $\{\rightarrow, \leftarrow, \backslash, /\}$

$$L(v) = \tilde{L}(\bar{v}),$$

bi-asymmetric operators $\{\perp, \mathrm{T}, \dashv, \vdash\}$

$$L(v) = \bar{L}(\bar{v}).$$

With respect to the rotation of logic coordinates the set of logic operations represents a symmetry group. Under the complete operation of symmetry the logic quantities in logic space are conserved. Among the symmetry transformations in logic vector space, special significance attaches to those transformations that leave the Euclidian norm of the vectors intact, the orthogonal transformations θ. The orthogonal transformation changes the orthonormal basis v into another orthonormal basis, the complementary basis $\bar{v}$. Instead of complementation of the basis we can equally well carry out the orthogonal transformation θ of a logic operator:

$$L(\bar{v}) \;=\; \theta L(v) \;=\; \rceil L(v)\rceil \;,$$

where $\rceil$ is a matrix negation operator.

Consequently, in logic space the combined operation of symmetry,

$$L(v) \;=\; \theta L(\bar{v}) \;,$$

is the identity transformation. A change-over from true to false, accompanied by the orthogonal transformation of an operator, leaves the logic operator intact.

With respect to logic operators the orthogonal transformation θ allows us to establish an elegant relation between transposition and converse transposition:

$$\overline{L} = \rceil \tilde{L} \rceil$$

$$\tilde{L} = \rceil \overline{L} \rceil$$

2.2 Discrete Universe $\Omega^{\square}$

The vector space formalism offers a concise, consistent notation which is ideally suited to the needs of matrix logic. With the appropriate rotation of logic coordinates we can always orientate logic vectors strictly along the verum or falsum reference axes. When this occurs, the logic vectors are simply the basis logic vectors. The logic vectors to be brought into the binary basis form later will be denoted by

$$|x\rangle,\ |y\rangle,\ |z\rangle,\ |t\rangle.$$

In the discrete universe we consider the binary-valued vectors of the unit norm

$$\langle x|x\rangle = \langle\bar{x}|\bar{x}\rangle = 1$$

with the inner product of two vectors defined as the Kronecker delta function:

$$\langle x|y\rangle = \delta_{xy} = \begin{cases} 1 & \text{if } x = y \\ 0 & \text{if } x \neq y \end{cases}$$

This implies that

$$\langle x|y\rangle \in E_2 = \{0,1\}.$$

Thus the discrete universe is a subspace of logic space inhabited just by the basis logic vectors

$$\langle 1| = (0\ \ 1)$$

$$\langle 0| = (1\ \ 0)$$

and the undecided vectors

$$\langle \mathbf{1}| = (1\ \ 1)$$

$$\langle \mathbf{0}| = (0\ \ 0)$$

which we shall discuss shortly.

The addition of logic vectors with the same truth status in the discrete logic $\Omega^{\square}$ is forbidden:

$$|x\rangle + |x\rangle \quad \text{— forbidden}$$

But

$$|x\rangle + |\bar{x}\rangle = |\mathbf{1}\rangle$$

$$|x\rangle - |x\rangle = |\mathbf{0}\rangle$$

are admissible.

2.2.1 Atomic transformations

The logic space is a space of interaction between logic states. The interaction between state $\langle x|$ and state $|y\rangle$ is activated when a logic operator L is inserted between the two adjoint vector spaces:

$$\langle x|L|y\rangle = x_1L_{11}y_1 + x_1L_{12}y_2 + x_2L_{21}y_1 + x_2L_{22}y_2$$

where L_{11}, L_{12}, L_{21}, L_{22} are matrix elements of a logic operator L, and x_1, x_2, y_1, y_2 are verum and falsum components of the bra and ket logic vectors $\langle x|$ and $|y\rangle$.
The fundamental bracket device $\langle x|L|y\rangle$, being inserted with a logic operator and a pair of logic vectors, will produce correct and meaningful results for all logic functions without exception.
Consider the following atomic transformations, which are due to the bracket products of conjunction $\langle x|\wedge|y\rangle$:

$$(1 \quad 0)\begin{pmatrix}0 & 0\\0 & 1\end{pmatrix}\begin{pmatrix}1\\0\end{pmatrix} = (0 \quad 0)\begin{pmatrix}1\\0\end{pmatrix} = (1 \quad 0)\begin{pmatrix}0\\0\end{pmatrix} = 0$$

$$(1 \quad 0)\begin{pmatrix}0 & 0\\0 & 1\end{pmatrix}\begin{pmatrix}0\\1\end{pmatrix} = (0 \quad 0)\begin{pmatrix}0\\1\end{pmatrix} = (1 \quad 0)\begin{pmatrix}0\\1\end{pmatrix} = 0$$

$$(0 \quad 1)\begin{pmatrix}0 & 0\\0 & 1\end{pmatrix}\begin{pmatrix}1\\0\end{pmatrix} = (0 \quad 1)\begin{pmatrix}1\\0\end{pmatrix} = (0 \quad 1)\begin{pmatrix}0\\0\end{pmatrix} = 0$$

$$(0 \quad 1)\begin{pmatrix}0 & 0\\0 & 1\end{pmatrix}\begin{pmatrix}0\\1\end{pmatrix} = (0 \quad 1)\begin{pmatrix}0\\1\end{pmatrix} = (0 \quad 1)\begin{pmatrix}0\\1\end{pmatrix} = 1$$

and disjunction <x | ∨ | y>:

$$(1\ \ 0)\begin{pmatrix}0 & 1\\ 1 & 1\end{pmatrix}\begin{pmatrix}1\\ 0\end{pmatrix} = (0\ \ 1)\begin{pmatrix}1\\ 0\end{pmatrix} = (1\ \ 0)\begin{pmatrix}0\\ 1\end{pmatrix} = 0$$

$$(1\ \ 0)\begin{pmatrix}0 & 1\\ 1 & 1\end{pmatrix}\begin{pmatrix}0\\ 1\end{pmatrix} = (0\ \ 1)\begin{pmatrix}0\\ 1\end{pmatrix} = (1\ \ 0)\begin{pmatrix}1\\ 1\end{pmatrix} = 1$$

$$(0\ \ 1)\begin{pmatrix}0 & 1\\ 1 & 1\end{pmatrix}\begin{pmatrix}1\\ 0\end{pmatrix} = (1\ \ 1)\begin{pmatrix}1\\ 0\end{pmatrix} = (0\ \ 1)\begin{pmatrix}0\\ 1\end{pmatrix} = 1$$

$$(0\ \ 1)\begin{pmatrix}0 & 1\\ 1 & 1\end{pmatrix}\begin{pmatrix}0\\ 1\end{pmatrix} = (1\ \ 1)\begin{pmatrix}0\\ 1\end{pmatrix} = (0\ \ 1)\begin{pmatrix}1\\ 1\end{pmatrix} = 1$$

Here it is opportune to emphasize a major advantage of matrix logic as compared to Boolean logic algebra. The algebraic mapping in the latter is restricted to just three logic connectives, $\{\neg, \wedge, \vee\}$. This is commonly viewed as an important advantage of Boolean logic. As a matter of fact, it rather points up its major defect, stemming from its inability to obtain a direct mathematical mapping for the remaining set of logic connectives. With the matrix operator formulation of logic, all logic connectives without exception obtain direct mathematical status.

Vectorization of truth-values

The interaction of bra and ket logic vectors in the atomic product results in the scalar truth-value α:

$$\alpha = \langle x | L | y \rangle$$

Because the obtained value α is the verum component of some logic vector, the corresponding logic vector can be easily resurrected with the support of the normalization condition:

$$| \alpha \rangle = \begin{pmatrix}\bar{\alpha}\\ \alpha\end{pmatrix}$$

This vectorization of truth-values is instrumental in the transition from conventional scalar logic to matrix logic.

Unary and binary operations

In conventional logic there is a strict division into unary and binary connectives. In matrix logic we have given the logic connectives the direct mathematical status of matrix operators, replacing the 16 binary Boolean operations with the universal procedure of matrix multiplication. It seems quite logical to consider, in addition to the binary operations, the unary operations of logic operators. In doing so, one ought to recognize that in matrix logic, as in conventional logic, a division into unary and binary operations does exist, but a division into unary and binary connectives or operators ceases to have similar implications. Indeed, in contrast to conventional connectives, the same logic operator is able to act either as a binary operator between two logic vectors, or as a unary operator upon a single logic vector. This implies, furthermore, that the set of 4 unary operations of Boolean logic is enlarged to a set of 16 unary operations, making the number of binary and unary operations in matrix logic equal, in total, 32 operations.

Equivalence and identity

The binary application of the equivalence operator in the matrix formulation of logic is the Kronecker function:

$$\langle x | \leftrightarrow | y \rangle = \langle x | y \rangle = \delta_{xy} = \begin{cases} 1 & \text{for } x = y \\ 0 & \text{for } x \neq y \end{cases}$$

On the other hand the equivalence operator, being applied as a unary operator, represents affirmation or the identity function I: clearly, multiplication of a vector by a unit matrix reproduces the given vector:

$$\leftrightarrow | x\rangle = I | x\rangle = | x\rangle ,$$

$$\langle x | \leftrightarrow = \langle x | I = \langle x | .$$

The negation operator

The analysis of the dual nature of matrix logic operators has finally brought us to the point where the reason for interpreting the inequivalence or strong alternation operator as the negation operator $\rceil$ can be meaningfully explained.

The binary matrix inequivalence operator, acting as a unary operator upon the logic vectors, converts them into complementary vectors. We have for the ket-logic vectors:

$$\begin{pmatrix} 0 & 1 \\ 1 & 0 \end{pmatrix}\begin{pmatrix} 1 \\ 0 \end{pmatrix} = \begin{pmatrix} 0 \\ 1 \end{pmatrix}$$

$$\begin{pmatrix} 0 & 1 \\ 1 & 0 \end{pmatrix}\begin{pmatrix} 0 \\ 1 \end{pmatrix} = \begin{pmatrix} 1 \\ 0 \end{pmatrix}$$

and for the bra-logic vectors:

$$(1 \quad 0)\begin{pmatrix} 0 & 1 \\ 1 & 0 \end{pmatrix} = (0 \quad 1)$$

$$(0 \quad 1)\begin{pmatrix} 0 & 1 \\ 1 & 0 \end{pmatrix} = (1 \quad 0)$$

In view of the assignments given to the logic vectors, the inequivalence operator demonstrates in a matrix formulation of logic the fundamental properties of negation. The negation operator is the matrix complement of the identity operator. It represents the only case of a classical connective being given a new interpretation; this adds yet another possible interpretation to a complex connective which has already been interpreted as inequivalence and exclusive disjunction. The role of the fundamental matrix negation operator in matrix logic formalism is indispensable.

Left and right logics

By interpreting logic connectives as matrix operators and introducing two adjoint spaces of logic vectors, we implicitly defined two sorts of logics: left logic and right logic. This new concept is due to the fundamental ability of matrix logic operators to act on logic vectors either from the right (right logic) or from the left (left logic), or both (left/right logic).
The set of matrix logic operators $\{\Omega^{+}\}$ decays into two subsets: one of them

$$\left\{\Omega^{+}_{\text{sym}}\right\} = \left\{\leftrightarrow, \rceil, \wedge, \vee, \uparrow, \downarrow, \hat{1}, \hat{0}\right\}$$

acts identically both in left and in right logics, whereas the other subset

$$\left\{\Omega^{+}_{\text{asym}}\right\} = \{\rightarrow, \leftarrow, \backslash, /, \bot, \top, \dashv, \vdash\}$$

acts differently in left logic and in right logic. As indicated by the subscripts, the subset $\left\{\Omega^{+}_{\text{sym}}\right\}$ consists of operators whose matrices are symmetric, and subset $\left\{\Omega^{+}_{\text{asym}}\right\}$ of operators whose matrices are asymmetric.

Virtual logic states

We intentionally began developing the formalism of matrix logic with no new premisses beyond those we accept in conventional binary Boolean logic. However, thanks to the application of the powerful mathematics of matrix calculus, matrix logic rapidly begins to unfold new and hitherto hidden aspects of logic.

The unary application of such operators as $\leftrightarrow$ and $\rceil$ (in both left and right logic), $\bot$ and $\top$ (in left logic only), $\dashv$ and $\vdash$ (in right logic only) does not violate the logic normalization condition. However, the unary application of such operators as

$$\{\wedge, \vee, \rightarrow, \leftarrow, \backslash, /, \uparrow, \downarrow, \bot, \top, \dashv, \vdash, \hat{1}, \hat{0}\}$$

may lead to the emergence of two undecided logic vectors:

$$\begin{pmatrix}0\\0\end{pmatrix} \quad \text{and} \quad \begin{pmatrix}1\\1\end{pmatrix},$$

for which the logic normalization condition is not satisfied.

The undecided vectors appear in the following cases:

$\bot$	—	$(0\ \ 1)\begin{pmatrix}0 & 0\\1 & 1\end{pmatrix} = (1\ \ 1)$
	—	$(1\ \ 0)\begin{pmatrix}0 & 0\\1 & 1\end{pmatrix} = (0\ \ 0)$
$\top$	—	$(0\ \ 1)\begin{pmatrix}1 & 1\\0 & 0\end{pmatrix} = (0\ \ 0)$
	—	$(1\ \ 0)\begin{pmatrix}1 & 1\\0 & 0\end{pmatrix} = (1\ \ 1)$

$\dashv$	$\begin{pmatrix}0&1\\0&1\end{pmatrix}\begin{pmatrix}0\\1\end{pmatrix}=\begin{pmatrix}1\\1\end{pmatrix}$ $\begin{pmatrix}0&1\\0&1\end{pmatrix}\begin{pmatrix}1\\0\end{pmatrix}=\begin{pmatrix}0\\0\end{pmatrix}$	— —
$\vdash$	$\begin{pmatrix}1&0\\1&0\end{pmatrix}\begin{pmatrix}0\\1\end{pmatrix}=\begin{pmatrix}0\\0\end{pmatrix}$ $\begin{pmatrix}1&0\\1&0\end{pmatrix}\begin{pmatrix}1\\0\end{pmatrix}=\begin{pmatrix}1\\1\end{pmatrix}$	— —
$\hat{1}$	$\begin{pmatrix}1&1\\1&1\end{pmatrix}\begin{pmatrix}0\\1\end{pmatrix}=\begin{pmatrix}1\\1\end{pmatrix}$ $\begin{pmatrix}1&1\\1&1\end{pmatrix}\begin{pmatrix}1\\0\end{pmatrix}=\begin{pmatrix}1\\1\end{pmatrix}$	$(0\ \ 1)\begin{pmatrix}1&1\\1&1\end{pmatrix}=(1\ \ 1)$ $(1\ \ 0)\begin{pmatrix}1&1\\1&1\end{pmatrix}=(1\ \ 1)$
$\hat{0}$	$\begin{pmatrix}0&0\\0&0\end{pmatrix}\begin{pmatrix}0\\1\end{pmatrix}=\begin{pmatrix}0\\0\end{pmatrix}$ $\begin{pmatrix}0&0\\0&0\end{pmatrix}\begin{pmatrix}1\\0\end{pmatrix}=\begin{pmatrix}0\\0\end{pmatrix}$	$(0\ \ 1)\begin{pmatrix}0&0\\0&0\end{pmatrix}=(0\ \ 0)$ $(1\ \ 0)\begin{pmatrix}0&0\\0&0\end{pmatrix}=(0\ \ 0)$
$\wedge$	— $\begin{pmatrix}0&0\\0&1\end{pmatrix}\begin{pmatrix}1\\0\end{pmatrix}=\begin{pmatrix}0\\0\end{pmatrix}$	— $(1\ \ 0)\begin{pmatrix}0&0\\0&1\end{pmatrix}=(0\ \ 0)$
$\vee$	$\begin{pmatrix}0&1\\1&1\end{pmatrix}\begin{pmatrix}0\\1\end{pmatrix}=\begin{pmatrix}1\\1\end{pmatrix}$ —	$(0\ \ 1)\begin{pmatrix}0&1\\1&1\end{pmatrix}=(1\ \ 1)$ —
$\uparrow$	$\begin{pmatrix}1&0\\0&0\end{pmatrix}\begin{pmatrix}0\\1\end{pmatrix}=\begin{pmatrix}0\\0\end{pmatrix}$ —	$(0\ \ 1)\begin{pmatrix}1&0\\0&0\end{pmatrix}=(0\ \ 0)$ —
$\downarrow$	— $\begin{pmatrix}1&1\\1&0\end{pmatrix}\begin{pmatrix}1\\0\end{pmatrix}=\begin{pmatrix}1\\1\end{pmatrix}$	— $(1\ \ 0)\begin{pmatrix}1&1\\1&0\end{pmatrix}=(1\ \ 1)$

$\rightarrow$	$\begin{pmatrix}1&1\\0&1\end{pmatrix}\begin{pmatrix}0\\1\end{pmatrix}=\begin{pmatrix}1\\1\end{pmatrix}$ —	— $(1\ \ 0)\begin{pmatrix}1&1\\0&1\end{pmatrix}=(1\ \ 1)$
$\leftarrow$	— $\begin{pmatrix}1&0\\1&1\end{pmatrix}\begin{pmatrix}1\\0\end{pmatrix}=\begin{pmatrix}1\\1\end{pmatrix}$	$(0\ \ 1)\begin{pmatrix}1&0\\1&1\end{pmatrix}=(1\ \ 1)$ —
$\backslash$	$\begin{pmatrix}0&0\\1&0\end{pmatrix}\begin{pmatrix}0\\1\end{pmatrix}=\begin{pmatrix}0\\0\end{pmatrix}$ —	— $(1\ \ 0)\begin{pmatrix}0&0\\1&0\end{pmatrix}=(0\ \ 0)$
$/$	— $\begin{pmatrix}0&1\\0&0\end{pmatrix}\begin{pmatrix}1\\0\end{pmatrix}=\begin{pmatrix}0\\0\end{pmatrix}$	$(0\ \ 1)\begin{pmatrix}0&1\\0&0\end{pmatrix}=(0\ \ 0)$ —

According to the analysis given above the only logic which operates strictly in terms of true or false, in both left and right logics, is the logic of matrix equivalence and matrix negation $\{\leftrightarrow, \rceil\}$.

Because in the bracket products <x | L | y> the undecided vectors appear as intermediary states of computation, which do not show themselves in the final results, they can be conveniently associated with the virtual logic states. Taking into account the virtual states, the logic space coordinate system will be altered as shown in Fig. 2.

The logic space is divided by the fundamental normalization square into two areas, internal and external, of virtual logic states. We have for the external area

$$1 \leq |\bar{p}| + |p| \leq 2 \ ,$$

and for the internal area

$$0 \leq |\bar{p}| + |p| \leq 1 \ .$$

This completes the definition of logic space which was begun in the previous section.

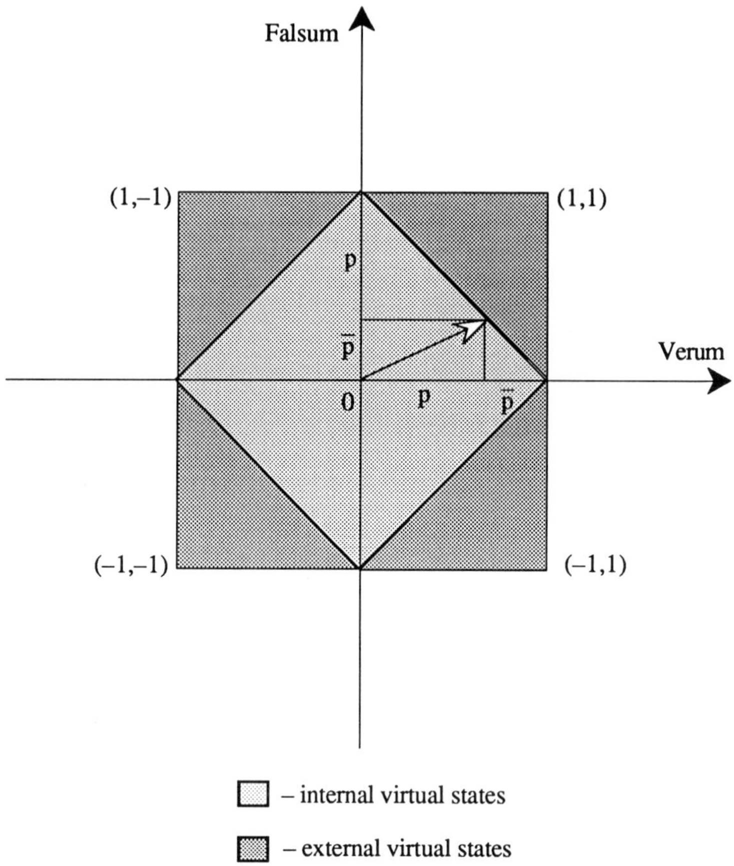

Fig. 2 Complete logic coordinate system

Logic norm

We define a norm μ for logic operators according to the formula:

$$\mu(L) = \| L \| = \sum_{i,j}^{2} L_{ij} \ .$$

The operator norm is the sum total of the elements of a matrix logic operator

$$\mu\,(L) = L_{11} + L_{12} + L_{21} + L_{22}\,.$$

The octet of symmetric logic operators $\{I, \rceil, \wedge, \vee, \uparrow, \downarrow, \hat{1}, \hat{0}\}$ forms a strong symmetric pattern with respect to the logic norm and orientation of logic space:

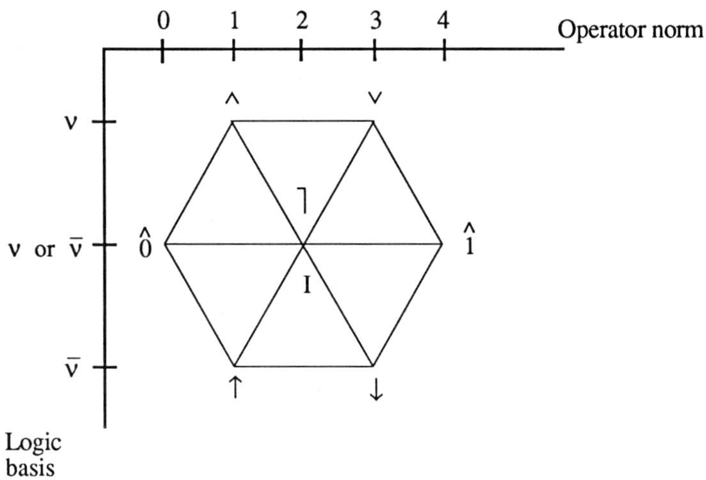

The contravertices in the octet are linked by complementation:

$$\bar{\wedge} = \downarrow$$
$$\bar{\uparrow} = \vee$$
$$\bar{\hat{0}} = \bar{\hat{1}}$$

which changes the norm of the operators accordingly. The operators $\rceil$ and I are placed in the same position in the centre, because complementation does not alter their norm:

$$\bar{\rceil} = I \qquad \text{but} \qquad \mu(\rceil) = \mu(I)\,.$$

The vertical links in the octet are due to the inversion of the logic space basis:

$$\wedge\,(v) = \uparrow(\bar{v})$$
$$\vee\,(v) = \downarrow(\bar{v})\,.$$

The horizontal links connect the dual logic operators:

$$\wedge^* = \vee \qquad \text{and} \qquad \uparrow^* = \downarrow$$

Multiplication table

In conventional logic compound expressions are formed from alternating logic variables and logic connectives. The only exception is the unary connective negation, which can be preceded by some other connective. All other situations when one connective is followed by another connective have no meaning and are rejected as not correctly formed formulas. However, in conventional logic the exceptional case of negation does not imply an interaction of the negation connective with the binary connective which precedes it: the negation first must be applied to a logic variable, and only then can a binary connective act between the result obtained and some other logic variable.
In marked contrast, matrix logic introduces a fundamental innovation. Here, a logic operator of negation, if placed immediately after another logic operator, acquires the meaning of matrix multiplication. Given the associativity of the latter, it can interact either with a logic vector or with the preceding operator, leading in both cases to the identical final result:

$$\langle x | L \left(\rceil | y \rangle \right) = \langle x | \left(L \rceil \right) | y \rangle$$

Furthermore, we can now generalize this case for any two logic operators, not necessarily involving the negation operator, and consider that any two operators, being brought together, acquire the meaning of matrix multiplication. The product is another logic operator which is still capable of interacting with another operator or acting upon a logic vector. The concept of the direct interaction of connectives, inconceivable in conventional logic, obtains in matrix logic a sound mathematical foundation and provides an important clue for understanding the higher level of abstraction. Whereas in conventional logic connectives act upon logic variables only, in the intelligent operations of the mind a logic operator can be modified by another operator, giving rise to logic operations of a higher level of abstraction.
Complete information about the interactions of logic operators is contained in the multiplication table of matrix logic given in Table 2.1.

To be consistent with the matrix logic approach, we must exclude all those interactions which violate the condition of binarity of the product. These forbidden interactions are indicated in the multiplication table by empty squares. Imposition of such a physical restriction implies that interaction of any two logic operators again produces a logic operator.
As a consequence of this constraint we must exclude from the 256 possible interactions 49 interactions as forbidden by matrix logic. The operational set of the latter thus consists of 207 multiplicative interactions.
The multiplication table clearly shows the special role played by the subset of identity and negation $\{\leftrightarrow, \rceil\}$: interacting with the set Ω^+, identity and negation uniquely map the latter on

×	↔	⌉	∧	∨	→	←	↓	↑	\	/	⊥	T	⊣	⊢	1	0
↔	↔	⌉	∧	∨	→	←	↓	↑	\	/	⊥	T	⊣	⊢	1	0
⌉	⌉	↔	/	→	∨	↓	←	\	↑	∧	T	⊥	⊣	⊢	1	0
∧	∧	\	∧	⊥	∧	⊥	\	0	\	0	⊥	0	∧	\	⊥	0
∨	∨	←	⊣					\	⊢	∧	1	⊥				0
→	→	↓	⊣					↑	⊢	/	1	T				0
←	←	∨	∧					⊢	\	⊣	⊥	1				0
↓	↓	→	/					⊢	↑	⊣	T	1				0
↑	↑	/	0	/	T	↑	T	↑	0	/	0	T	/	↑	T	0
\	\	∧	0	∧	⊥	\	⊥	\	0	∧	0	⊥	∧	\	⊥	0
/	/	↑	/	T	/	T	↑	0	↑	0	T	0	/	↑	T	0
⊥	⊥	⊥	∧					\	\	∧	⊥	⊥				0
T	T	T	/					↑	↑	/	T	T				0
⊣	⊣	⊢	⊣	1	⊣	1	⊢	0	⊢	0	1	0	⊣	⊢	1	0
⊢	⊢	⊣	0	⊣	1	⊢	1	⊢	0	⊣	0	1	⊣	⊢	1	0
1	1	1	⊣					⊢	⊢	⊣	1	1				0
0	0	0	0	0	0	0	0	0	0	0	0	0	0	0	0	0

Table 2.1 Multiplication table of matrix logic

itself. Apart from interactions with {↔, ⌉}, interactions of the remaining logic operators with the set Ω^+ reduce the latter to a decuplet subset consisting of the following 10 operators:

$$\{\wedge, /, \backslash, \uparrow, \perp, \mathrm{T}, \dashv, \vdash, \hat{1}, \hat{0}\}$$

A great deal of useful information concerning the behaviour and properties of logic operators can be deduced from the multiplication table of matrix logic.

Cancellation rule

In addition to the forbidden interactions several other important properties distinguish matrix logic algebra from ordinary logic algebra.

In ordinary algebra it follows from

$$AB = CB ,$$

due to the cancellation rule, that

$$A = C.$$

However, in matrix algebra

$$L_1L_2 = L_3L_2$$

does not necessarily imply that

$$L_1 = L_3$$

Consider

$$\rightarrow / = \uparrow / \quad \text{but} \quad \rightarrow \neq \uparrow$$

or

$$\uparrow \wedge = \backslash \wedge \quad \text{but} \quad \uparrow \neq \backslash.$$

Thus, the cancellation rule of ordinary algebra is not universally applicable in matrix logic. Multiplicity is an essential characteristic of high-level intelligence.

Null product

Many interactions in the multiplication table result in a null matrix. The null matrix $\hat{0}$ plays in matrix logic a role similar to that of the number 0 in ordinary algebra. However, contrary to the scalar case, where $AB = 0$ implies strictly that $A = 0$ or/and $B = 0$, the product of two matrices can be empty even if both of them differ from a null matrix. In the multiplication table we find 18 different cases of such orthogonal products:

$$\begin{array}{lll}
\wedge \uparrow, & \wedge /, & \wedge \top \\
\uparrow \wedge, & \uparrow \backslash, & \uparrow \bot \\
\backslash \wedge, & \backslash \backslash, & \backslash \bot \\
/ \uparrow, & / /, & / \top \\
\dashv \uparrow, & \dashv /, & \dashv \top \\
\vdash \wedge, & \vdash \backslash, & \vdash \dashv
\end{array}$$

Particular importance attaches to the nullifying products of the projective operators $\wedge$ and $\uparrow$:

$$\begin{pmatrix} 0 & 0 \\ 0 & 1 \end{pmatrix}\begin{pmatrix} 1 & 0 \\ 0 & 0 \end{pmatrix} = \begin{pmatrix} 0 & 0 \\ 0 & 0 \end{pmatrix},$$

and to the self-nullifying operators \ and / :

$$\begin{pmatrix} 0 & 0 \\ 1 & 0 \end{pmatrix}\begin{pmatrix} 0 & 0 \\ 1 & 0 \end{pmatrix} = \begin{pmatrix} 0 & 0 \\ 0 & 0 \end{pmatrix},$$

$$\begin{pmatrix} 0 & 1 \\ 0 & 0 \end{pmatrix}\begin{pmatrix} 0 & 1 \\ 0 & 0 \end{pmatrix} = \begin{pmatrix} 0 & 0 \\ 0 & 0 \end{pmatrix}.$$

Commutation

Finally, one of the fundamental features of matrix logic concerns the non-commutativity of operations. Whereas in ordinary algebra

$$AB = BA,$$

the operational logic set consists mostly of non-commuting operators. This fact has profound implications for the fundamental theory of the mechanism of the mind. Non-commutation algebra is a distinct feature of high-level intelligence where the commutation relations impose fundamental limits on logic inference and observation.

Transmultiplication

Useful data are also given in the transmultiplication table of logic operators, which is defined as follows. If L' and L'' are logic operators

$$L' = \begin{pmatrix} L'_{11} & L'_{12} \\ L'_{21} & L'_{22} \end{pmatrix}; \quad L'' = \begin{pmatrix} L''_{11} & L''_{12} \\ L''_{21} & L''_{22} \end{pmatrix},$$

then their transmultiplication product is a component-by-component multiplication matrix:

$$L' \ast L'' = \begin{pmatrix} L'_{11}L''_{11} & L'_{12}L''_{12} \\ L'_{21}L''_{21} & L'_{22}L''_{22} \end{pmatrix}.$$

The radiating dot ✸ will be used in order to distinguish transmultiplication from the ordinary matrix product, denoted by juxtaposition.

Clearly, the transmultiplication table is symmetric with respect to the main diagonal, whereas in the multiplication table the upper and lower sections are transposes of each other.

✸	↔	⌉	∧	∨	→	←	↓	↑	\	/	⊥	T	⊣	⊢	**1**	0
↔	↔	●	∧	∧	↔	↔	↑	↑	0	0	∧	↑	∧	↑	↔	0
⌉	●	⌉	0	⌉	/	\	⌉	0	\	/	\	/	/	\	⌉	0
∧	∧	0	∧	∧	∧	∧	●	0	0	0	∧	0	∧	0	∧	0
∨	∧	⌉	∧	∨	⊣	⊥	⌉	●	\	/	⊥	/	⊣	\	∨	0
→	↔	/	∧	⊣	→	↔	T	↑	●	/	∧	T	⊣	↑	→	0
←	↔	\	∧	⊥	↔	←	⊢	↑	\	●	⊥	↑	∧	⊢	←	0
↓	↔	⌉	●	⌉	T	⊢	↓	↑	\	/	\	T	/	⊢	↓	0
↑	↑	0	0	●	↑	↑	↑	↑	0	0	0	↑	0	↑	↑	0
\	0	\	0	\	●	\	\	0	\	0	\	0	0	\	\	0
/	0	/	0	/	/	●	/	0	0	/	0	/	/	0	/	0
⊥	∧	\	∧	⊥	∧	⊥	\	0	\	0	⊥	●	∧	\	⊥	0
T	↑	/	0	/	T	↑	T	↑	0	/	●	T	/	↑	T	0
⊣	∧	/	∧	⊣	⊣	∧	/	0	0	/	∧	/	⊣	●	⊣	0
⊢	↑	\	0	\	↑	⊢	⊢	↑	\	0	\	↑	●	⊢	⊢	0
1	↔	⌉	∧	∨	→	←	↓	↑	\	/	⊥	T	⊣	⊢	**1**	●
0	0	0	0	0	0	0	0	0	0	0	0	0	0	0	●	0

Table 2.2 Transmultiplication table of matrix logic

<u>Major basis set</u>

In conventional logic any logic formula can be expressed by the major basis set of connectives {⌉, ∧, ∨}. As follows from the multiplication table, the same holds true in matrix logic for the corresponding logic operators. Explicitly, we have the following equalities:

$\rceil = \rceil$	$\uparrow = \rceil \wedge \rceil$
$\wedge = \wedge$	$\downarrow = \rceil \vee \rceil$
$\vee = \vee$	$\perp = \wedge \vee$
$\leftrightarrow = \rceil \rceil$	$\top = \rceil \wedge \vee \rceil$
$\rightarrow = \rceil \vee$	$\dashv = \vee \wedge$
$\leftarrow = \vee \rceil$	$\vdash = \rceil \vee \wedge \rceil$
$\backslash = \wedge \rceil$	$\hat{1} = \vee \wedge \vee$
$/ = \rceil \wedge$	$\hat{0} = \wedge \rceil \wedge$

We have generated the complete set of logic operators Ω^+ from three basic operators:

$$\{\rceil, \wedge, \vee\}\ .$$

Complement operator

The matrix logic complement operation can be formalized in the following relation:

$$\overline{\alpha} = \mathbf{1} - \alpha$$

where α and $\overline{\alpha}$ are the given matrix and its complement respectively, **1** is a corresponding unit constant, which becomes a number 1 if α is a scalar; a vector (1,1) if α is a logic vector; and the matrix $\begin{pmatrix} 1 & 1 \\ 1 & 1 \end{pmatrix}$ if α is a logic operator.

In the previous section we emphasized the special role of the matrix negation operator $\rceil$, which incorporates in itself the properties of exclusive disjunction, and inequivalence as well. The negation operator $\rceil$, acting upon a logical vector, produces a complementary vector:

$$\rceil | 0\rangle = | 1\rangle$$

$$\rceil | 1\rangle = | 0\rangle$$

where $| 0\rangle$, $| 1\rangle$ abbreviate (1,0) and (0,1) respectively.

Thus, in case of logic vectors, the operator $\rceil$ in fact acts as a complement operator.

The important relation which unifies the three different aspects of the complement operator can be proved in matrix logic. Namely, a complement of a bracket product is equal to a bracket product of the complementary operator:

$$\overline{<x|L|y>} = <x|\bar{L}|y> .$$

Indeed

$$1 - <x|L|y> = <x|\hat{1} - L|y> =$$
$$= <x\hat{1}|y> - <x|L|y> = 1 - <x|L|y>.$$

The important property of the complement operator consists in its ability to act in one of three different ways: either upon a scalar or upon an operator whether inside or outside the brackets. In dealing with compound logic expressions we can choose the most useful way. Independently of the way in which the complement operator is applied, the resulting expressions are logically equivalent and yield the same numerical values:

$$L_1 \overline{<x|L|y>} L_2 = L_1 <x|\bar{L}|y> L_2 =$$
$$= L_1 \rceil <x|L|y> L_2 =$$
$$= L_1 <x|L|y> \rceil L_2 .$$

The properties of the complement operator make it a useful tool for the transformation of logic expressions. It is important to bear in mind not only the similarity but also the difference in the application of the complement operator to a logic scalar, logic vector and logic operator:

$$\bar{x} = 1 - x ,$$
$$|\bar{x}> = \mathbf{1} - |x> = \rceil |x> ,$$
$$\bar{L} = \hat{1} - L \neq \rceil L .$$

Whereas complementation of a logic vector is equal to a product of the negation operator with the vector, this is not the case with the logic operators. The application of the complement operation $\bar{L} = \hat{1} - L$ divides the set Ω^+ into two complementary octet subsets:

$$\{\leftrightarrow\ ,\ \urcorner\}$$
$$\{\wedge\ ,\ \downarrow\}$$
$$\{\vee\ ,\ \uparrow\}$$
$$\{\rightarrow\ ,\ \backslash\}$$
$$\{\leftarrow\ ,\ /\}$$
$$\{\perp\ ,\ \top\}$$
$$\{\dashv\ ,\ \vdash\}$$
$$\{\hat{1}\ ,\ \hat{0}\}$$

The only thing that matrix negation and the complement operator have in common is the way they act upon logic vectors. But there the similarity ends. Matrix negation acts on each logic operator in different ways. However, for some operators the result can coincide with the complementation of an operator:

Left products:

$$\urcorner I = \bar{I}$$
$$\urcorner \urcorner = \bar{\urcorner}$$
$$\urcorner \perp = \bar{\perp}$$
$$\urcorner \top = \bar{\top}$$

Right products:

$$I \urcorner = \bar{I}$$
$$\urcorner \urcorner = \bar{\urcorner}$$
$$\dashv \urcorner = \bar{\dashv}$$
$$\vdash \urcorner = \bar{\vdash}$$

The complement of an operator-vector product is equal to the product with the complementary operator:

$$\overline{L|x>} = \bar{L}|x> .$$

Similarly, in bra vector space:

$$\overline{<x|L} = <x|\bar{L} .$$

Indeed,

$$\bar{L}|x> = (\hat{1} - L)|x> = \hat{1}|x> - L|x> = \overline{L|x>} .$$

Clearly

$$\overline{\overline{L}} = \overline{\hat{1} - L} = L\,.$$

Duality transformation

In conventional logic the major basis set $\{\neg, \wedge, \vee\}$ can be reduced to the minor basis sets:

$$\{\neg, \wedge\} \quad \text{or} \quad \{\neg, \vee\}.$$

This reduction is possible because of the De Morgan duality rules:

$$\overline{x \wedge y} = \bar{x} \vee \bar{y}$$

$$\overline{x \vee y} = \bar{x} \wedge \bar{y}$$

Introduction of the complement operator, as proposed in the previous section, makes it possible to formulate a matrix equivalent of the De Morgan rules, which reveals new fundamental properties of these rules and leads to a formulation of the generalized De Morgan rule.
Taking into account that the negation operator is the orthogonal operator,

$$\tilde{\neg} = \neg^{-1} = \neg,$$

the duality rules of conventional logic take in matrix logic the form of the orthogonal transformation, as the following derivations demonstrate:

$$\overline{<x|\wedge|y>} \;=\; <x|\downarrow|y> \;=\; <x|\neg\vee\neg|y> \;=\; <\bar{x}|\vee|\bar{y}>$$

$$\overline{<x|\vee|y>} \;=\; <x|\uparrow|y> \;=\; <x|\neg\wedge\neg|y> \;=\; <\bar{x}|\wedge|\bar{y}>\,.$$

Thus, reduction of the major basis set to the minor basis sets in matrix logic is achieved by the orthogonal transformation of conjunction or disjunction.
Clearly, the orthogonal transformation can be considered not only for the operators of conjunction and disjunction but for any logic operator. In this way we arrive at the generalized duality rule, where three distinct classes of the orthogonal transformation θ of logic operators can be specified:

$\theta L = \neg L \neg = L$ – invariant product

$\theta L = \neg L \neg = \bar{L}$ – complementary product

$\theta L = \neg L \neg = L' \neq L \neq \bar{L}$ – not invariant, not complementary

The following chart presents the products of orthogonal and anti-orthogonal transformations in the set of logic operators, which is closed under both operations:

L	θL	$\overline{\theta L}$
↔	↔	¬
¬	¬	↔
∧	↑	∨
∨	↓	∧
→	←	/
←	→	\
\	/	←
/	\	→
↑	∧	↓
↓	∨	↑
⊥	T	⊥
T	⊥	T
⊣	⊢	⊣
⊢	⊣	⊢
$\hat{1}$	$\hat{1}$	$\hat{0}$
$\hat{0}$	$\hat{0}$	$\hat{1}$

Table 2.3 Orthogonal and anti-orthogonal transforms in matrix logic.

The anti-orthogonal transformation of logic operators is a formalization of the empirical notion of dual connectives in conventional logic:

$$L^* = \overline{\theta L} \quad \text{and} \quad L = \overline{\theta L^*},$$

where L is a logic operator and L^* is its dual.

Then:

$$< x \mid L \mid y > \; = \; < x \mid \overline{\theta L^*} \mid y > \; =$$

$$= \; \overline{< x \mid \theta L^* \mid y >} \; .$$

Complementing both sides of the equation and exploring the associativity of matrix multiplication, we obtain the general duality rule:

$$\begin{aligned}\overline{\langle x | L | y\rangle} &= \langle x | \theta L^* | y\rangle = \\ &= \langle x | \neg L^* \neg | y\rangle = \\ &= \langle \bar{x} | L^* | \bar{y}\rangle .\end{aligned}$$

The three different responses of logic operators to orthogonal transformation give us three subsets of duality rules:

<u>Subset $L^* = \bar{L}$</u>

1. $\overline{\langle x | \leftrightarrow | y\rangle} = \langle x | \neg | y\rangle = \langle \bar{x} | \neg | \bar{y}\rangle$
2. $\overline{\langle x | \neg | y\rangle} = \langle x | \leftrightarrow | y\rangle = \langle \bar{x} | \leftrightarrow | \bar{y}\rangle$
3. $\overline{\langle x | \hat{1} | y\rangle} = \langle x | \hat{0} | y\rangle = \langle \bar{x} | \hat{0} | \bar{y}\rangle$
4. $\overline{\langle x | \hat{0} | y\rangle} = \langle x | \hat{1} | y\rangle = \langle \bar{x} | \hat{1} | \bar{y}\rangle$

<u>Subset $L^* = L$</u>

1. $\overline{\langle x | \perp | y\rangle} = \langle x | \top | y\rangle = \langle \bar{x} | \perp | \bar{y}\rangle$
2. $\overline{\langle x | \top | y\rangle} = \langle x | \perp | y\rangle = \langle \bar{x} | \top | \bar{y}\rangle$
3. $\overline{\langle x | \dashv | y\rangle} = \langle x | \vdash | y\rangle = \langle \bar{x} | \dashv | \bar{y}\rangle$
4. $\overline{\langle x | \vdash | y\rangle} = \langle x | \dashv | y\rangle = \langle \bar{x} | \vdash | \bar{y}\rangle$

<u>Subset $L^* \neq \bar{L} \neq L$</u>

1. $\overline{\langle x | \wedge | y\rangle} = \langle x | \downarrow | y\rangle = \langle \bar{x} | \vee | \bar{y}\rangle$
2. $\overline{\langle x | \vee | y\rangle} = \langle x | \uparrow | y\rangle = \langle \bar{x} | \wedge | \bar{y}\rangle$
3. $\overline{\langle x | \rightarrow | y\rangle} = \langle x | \backslash | y\rangle = \langle \bar{x} | / | \bar{y}\rangle$
4. $\overline{\langle x | \leftarrow | y\rangle} = \langle x | / | y\rangle = \langle \bar{x} | \backslash | \bar{y}\rangle$
5. $\overline{\langle x | \uparrow | y\rangle} = \langle x | \vee | y\rangle = \langle \bar{x} | \downarrow | \bar{y}\rangle$
6. $\overline{\langle x | \downarrow | y\rangle} = \langle x | \wedge | y\rangle = \langle \bar{x} | \uparrow | \bar{y}\rangle$
7. $\overline{\langle x | \backslash | y\rangle} = \langle x | \rightarrow | y\rangle = \langle \bar{x} | \leftarrow | \bar{y}\rangle$
8. $\overline{\langle x | / | y\rangle} = \langle x | \leftarrow | y\rangle = \langle \bar{x} | \rightarrow | \bar{y}\rangle$

Operations of symmetry

In the matrix formulation of logic the duality transformation represents the combined operation of symmetry in logic space. The orthogonal transformation of a logic operator is equivalent to the transition from the basis ν to the complementary basis $\bar{\nu}$, which recombines the falsum and verum components. Whenever this operation of symmetry is applied jointly with the operation of complementation to a dual logic operator, we obtain the fundamental equality:

$$< x \mid L \mid y > \;=\; < \bar{x} \mid \overline{L^*} \mid \bar{y} > .$$

The two operations of symmetry are precisely what is required to undo the operations of the orthogonal transformation and complementation that we employ to translate an operator into a dual operator.

Matrix operations provide many other ways of deriving duality rules. Consider, for instance, derivation through the decomposition of tautology:

$$\hat{1} = L + \bar{L}$$

Then we can write

$$<x \mid L \mid y> + <x \mid \bar{L} \mid y> \;=\; <x \mid \hat{1} \mid y> .$$

The value of tautology is 1:

$$<x \mid L \mid y> + <x \mid \bar{L} \mid y> \;=\; 1$$

$$\begin{aligned} <x \mid L \mid y> &= 1 - <x \mid \bar{L} \mid y> = \\ &= \overline{< x \mid \bar{L} \mid y >} = \\ &= \overline{< x \mid \theta^2 \bar{L} \mid y >} = \\ &= \overline{< \bar{x} \mid \theta \bar{L} \mid \bar{y} >} = \\ &= \overline{< \bar{x} \mid L^* \mid \bar{y} >} \qquad \blacksquare \end{aligned}$$

Derivation due simply to the properties of the complement operator:

$$\begin{aligned} < x \mid L \mid y > &= < x \mid \bar{\bar{L}} \mid y > = \\ &= \overline{< x \mid \bar{L} \mid y >} = \end{aligned}$$

$$= \overline{< x | \theta^2 \bar{L} | y >} =$$

$$= \overline{< \bar{x} | L^* | \bar{y} >} \qquad \blacksquare$$

The classical De Morgan rules for conjunction and disjunction will in this case be derived as follows:

$$< x | \wedge | y > = \overline{< x | \downarrow | y >} = \overline{< x | \theta^2 \downarrow | y >} = \overline{< \bar{x} | \vee | \bar{y} >}$$

$$< x | \vee | y > = \overline{< x | \uparrow | y >} = \overline{< x | \theta^2 \uparrow | y >} = \overline{< \bar{x} | \wedge | \bar{y} >} .$$

The material in this section illustrates the new deductive power of matrix logic. Incorporating classical logic as a special case of scalar relaxation and being in full agreement with its results, matrix logic at the same time delivers new results and generalizations. As a formalism with direct computational power, it allows us to uncover those aspects of logic which remain undetected and cannot be detected within the framework of conventional logic.

Transformation rules

By converting logical relations into matrix operator equations we are not merely introducing new symbolic notations, but devising a formalism in which the whole mechanics of reasoning becomes expressible in the language of matrix operator operations. We are now capable not only of transforming logic formulas by traditional logical means but also of employing the operations of matrix algebra. As soon as an operator equation is written down, we do not need to bother with what it actually represents in a framework of logical interpretation and can deal with it on a purely formal mathematical basis. The significance of this fact reveals itself in the insight that the transformation rules of conventional logic, the identification of which had somewhat the character of a discovery so that they even were assigned the name of the discoverer, like the De Morgan rules, the Switcheroo rule, or the Shiefer stroke, now can be deduced on purely formal grounds and without any logical evaluation whatsoever. Furthermore, the mathematically correct transformation of operator equations will produce the complete set of possible logic relations, including those which are not common in conventional logic.

Transposition rule

Matrix transposition formalizes the concept of converse logic functions and can be employed as a new technique of inference in matrix logic. For an atomic bracket product we make use of the fact that the transposition of a scalar is an invariant. Then we obtain

$$\langle x | L | y \rangle \;=\; \widetilde{\langle x | L | y \rangle} \;=\; |\tilde{y}\rangle \, \tilde{L} \, \langle \tilde{x}| \;=\; \langle y | \tilde{L} | x \rangle \;,$$

where $\tilde{L}$ is a converse logic operator.

Clearly, when L is a symmetric matrix, a logic function and converse logic function coincide:

$$\langle x | L | y \rangle = \langle x | \tilde{L} | y \rangle = \langle y | L | x \rangle \,.$$

This establishes a division of the set of logic functions into two octet subsets:

Symmetric function	Asymmetric function
$\langle x \vert \tilde{\leftrightarrow} \vert y \rangle = \langle x \vert \leftrightarrow \vert y \rangle$	$\langle x \vert \tilde{\rightarrow} \vert y \rangle = \langle x \vert \leftarrow \vert y \rangle$
$\langle x \vert \tilde{\rceil} \vert y \rangle = \langle x \vert \rceil \vert y \rangle$	$\langle x \vert \tilde{\leftarrow} \vert y \rangle = \langle x \vert \rightarrow \vert y \rangle$
$\langle x \vert \tilde{\wedge} \vert y \rangle = \langle x \vert \wedge \vert y \rangle$	$\langle x \vert \tilde{\setminus} \vert y \rangle = \langle x \vert / \vert y \rangle$
$\langle x \vert \tilde{\vee} \vert y \rangle = \langle x \vert \vee \vert y \rangle$	$\langle x \vert \tilde{/} \vert y \rangle = \langle x \vert \setminus \vert y \rangle$
$\langle x \vert \tilde{\uparrow} \vert y \rangle = \langle x \vert \uparrow \vert y \rangle$	$\langle x \vert \tilde{\perp} \vert y \rangle = \langle x \vert \dashv \vert y \rangle$
$\langle x \vert \tilde{\downarrow} \vert y \rangle = \langle x \vert \downarrow \vert y \rangle$	$\langle x \vert \tilde{T} \vert y \rangle = \langle x \vert \vdash \vert y \rangle$
$\langle x \vert \tilde{\hat{1}} \vert y \rangle = \langle x \vert \hat{1} \vert y \rangle$	$\langle x \vert \tilde{\dashv} \vert y \rangle = \langle x \vert \perp \vert y \rangle$
$\langle x \vert \tilde{\hat{0}} \vert y \rangle = \langle x \vert \hat{0} \vert y \rangle$	$\langle x \vert \tilde{\vdash} \vert y \rangle = \langle x \vert T \vert y \rangle$

Contraposition rule

Converse transposition, that is, transposition across the secondary matrix diagonal, embodies the important logic operation of contraposition. Again the complete logic set is divided into two octet subsets, one of which changes under converse transposition,

$$\{\wedge, \vee, \uparrow, \downarrow, \perp, T, \dashv, \vdash\} \;,$$

and another which remains invariant,

$$\{\leftrightarrow, \rceil, \rightarrow, \leftarrow, \setminus, /, \hat{1}, \hat{0}\} \;.$$

The first octet comprises symmetric and bi-asymmetric functions, the second octet comprises contrasymmetric and bisymmetric functions:

Symmetric and bi-asymmetric functions:

$$\langle x | \overset{\sqcap}{\wedge} | y\rangle = \langle x | \uparrow | y\rangle = \langle \bar{y} | \wedge | \bar{x}\rangle$$

$$\langle x | \overset{\sqcap}{\vee} | y\rangle = \langle x | \downarrow | y\rangle = \langle \bar{y} | \vee | \bar{x}\rangle$$

$$\langle x | \overset{\sqcap}{\downarrow} | y\rangle = \langle x | \vee | y\rangle = \langle \bar{y} | \downarrow | \bar{x}\rangle$$

$$\langle x | \overset{\sqcap}{\uparrow} | y\rangle = \langle x | \wedge | y\rangle = \langle \bar{y} | \uparrow | \bar{x}\rangle$$

$$\langle x | \overset{\sqcap}{\perp} | y\rangle = \langle x | \vdash | y\rangle = \langle \bar{y} | \perp | \bar{x}\rangle$$

$$\langle x | \overset{\sqcap}{\top} | y\rangle = \langle x | \dashv | y\rangle = \langle \bar{y} | \top | \bar{x}\rangle$$

$$\langle x | \overset{\sqcap}{\dashv} | y\rangle = \langle x | \top | y\rangle = \langle \bar{y} | \dashv | \bar{x}\rangle$$

$$\langle x | \overset{\sqcap}{\vdash} | y\rangle = \langle x | \perp | y\rangle = \langle \bar{y} | \vdash | \bar{x}\rangle$$

Contrasymmetric and bisymmetric functions:

$$\langle x | \overset{\sqcap}{\rightarrow} | y\rangle = \langle x | \rightarrow | y\rangle = \langle \bar{y} | \rightarrow | \bar{x}\rangle$$

$$\langle x | \overset{\sqcap}{\leftarrow} | y\rangle = \langle x | \leftarrow | y\rangle = \langle \bar{y} | \leftarrow | \bar{x}\rangle$$

$$\langle x | \overset{\sqcap}{\backslash} | y\rangle = \langle x | \backslash | y\rangle = \langle \bar{y} | \backslash | \bar{x}\rangle$$

$$\langle x | \overset{\sqcap}{/} | y\rangle = \langle x | / | y\rangle = \langle \bar{y} | / | \bar{x}\rangle$$

$$\langle x | \overset{\sqcap}{\leftrightarrow} | y\rangle = \langle x | \leftrightarrow | y\rangle = \langle \bar{y} | \leftrightarrow | \bar{x}\rangle$$

$$\langle x | \overset{\sqcap}{\rceil} | y\rangle = \langle x | \rceil | y\rangle = \langle \bar{y} | \rceil | \bar{x}\rangle$$

$$\langle x | \overset{\sqcap}{\hat{1}} | y\rangle = \langle x | \hat{1} | y\rangle = \langle \bar{y} | \hat{1} | \bar{x}\rangle$$

$$\langle x | \overset{\sqcap}{\hat{0}} | y\rangle = \langle x | \hat{0} | y\rangle = \langle \bar{y} | \hat{0} | \bar{x}\rangle$$

With respect to the contraposition form, we establish four different conservation laws:

Bisymmetric operators $\{I, \rceil, \hat{1}, \hat{0}\}$:

$$<x|L|y> = <\bar{y}|L|\bar{x}> ,$$

Contrasymmetric operators $\{\rightarrow, \leftarrow, \backslash, /\}$:

$$<x|L|y> = <\bar{y}|\tilde{L}|\bar{x}> ,$$

Symmetric operators $\{\wedge, \vee, \downarrow, \uparrow\}$:

$$<x|L|y> = <\bar{y}|\overline{L}|\bar{x}> ,$$

Bi-asymmetric operators $\{\perp, \top, \dashv, \vdash\}$:

$$<x|L|y> = <\bar{y}|\bar{L}|\bar{x}> .$$

In matrix logic the inference of one set of formulas from another set takes the form of rigorous mathematical derivations.

Identity decomposition rule

The identity operator can always be inserted at any place into a matrix logic expression without causing any change in the latter. However, if the identity is first presented as a square product of the negation operator,

$$I = \rceil^2 ,$$

and is then inserted between a logic operator and a logic vector, it modifies the logic vector into its complement and the logic operator according to the multiplication table:

$$<x|L|y> = <x|\rceil^2 L|y> = <\bar{x}|\rceil L|y> .$$

Inserting the identity operator at another position, we obtain another formula for the same logic function:

$$<x|L|y> = <x|L\rceil^2|y> = <x|L\rceil|\bar{y}> .$$

Through a bilateral application of the identity decomposition rule we can transfer negation from the antecedent to the consequent and vice versa:

$$<\bar{x}|L|y> = <\bar{x}|\rceil^2 L\rceil^2|y> = <\bar{x}|\theta^2 L|y> = <x|\theta L|\bar{y}> .$$

The latter transformation involves the square of the orthogonal transformation θ^2.

Orthogonal transformation rule and complementation rule

We have become acquainted with these rules of matrix logic in the preceding sections. In general form they are written as follows:

$$<x|L|y> \;=\; <\bar{x}|\theta L|\bar{y}>,$$

$$<x|L|y> \;=\; \overline{<x|\bar{L}|y>}.$$

The orthogonal transformation rule actually can be derived in a purely formal way from the identity decomposition rule. The complementation rule follows directly from the definition of the complement operator. The latter rule allows us to extract the negation (complement) operator from an operator and either to apply it to the complete logic expression or to multiply it with the first neighbouring operator.

Combined transformation, involving both the orthogonal transformation and the complementation rules, automatically leads to the general duality transform rule:

$$<x|L|y> = <\bar{x}|\theta L|\bar{y}> = \overline{<\bar{x}|\overline{\theta L}|\bar{y}>}\,.$$

General covariancy rule

Any atomic logical interaction is invariant with respect to the following fundamental operations of symmetry, executed jointly: orthogonal transformation, transposition, and contraposition:

$$<x|L|y> \;=\; <x|\theta\,\tilde{\bar{L}}\,|y>\,.$$

This is a consequence of the fact that the orthogonal transformation is equivalent to the inversion of the basis of logic space:

$$<x|\theta\,\tilde{\bar{L}}\,|y> \;=\; <\bar{x}|\,\tilde{\bar{L}}\,|\bar{y}>\,,$$

whereas the combined operation of transposition and contraposition nullifies the induced inversion:

$$<\bar{x}|\,\tilde{\bar{L}}\,|\bar{y}> \;=\; <\bar{\bar{x}}|L|\bar{\bar{y}}> \;=\; <x|L|y>\,.$$

The covariance rule has a profound logic meaning. The structure of any logic operator is conserved whenever we (1) rotate logic space to obtain a complementary basis $\overline{v}$, that is, recombine verum and falsum components; (2) recombine cause (antecedent) and effect (consequent); (3) perform contraposition. The fundamental operations of symmetry commute and the order of their execution is therefore irrelevant.

The change-over from true to false has no impact on bisymmetric logic functions:

$$<x \mid L \mid y> \;=\; <\overline{x} \mid L \mid \overline{y}> .$$

Recombination of cause and effect has no impact on symmetric functions:

$$<x \mid L \mid y> \;=\; <y \mid L \mid x> .$$

Both operations of symmetry applied jointly have no impact on contrasymmetric functions:

$$<x \mid L \mid y> \;=\; <\overline{y} \mid L \mid \overline{x}> .$$

Finally, for bi-asymmetric filters the combined operation of symmetry is a triple which, in addition to the two operations discussed, also requires the converse transposition of the operator:

$$<x \mid L \mid y> \;=\; <\overline{y} \mid \overset{\sqcap}{L} \mid \overline{x}> .$$

The contraposition rule, playing an important role in logic deduction, can be derived from the general covariance rule:

$$<x \mid L \mid y> \;=\; <x \mid \theta\, \overset{\sim}{\overset{\sqcap}{L}} \mid y> \;=\; <x \mid \theta\, \overset{\sqcap}{\tilde{L}} \mid y> \;=\; <\overline{x} \mid \overset{\sqcap}{\tilde{L}} \mid \overline{y}> .$$

From this follows:

$$<x \mid \overset{\sqcap}{L} \mid y> \;=\; <\overline{x} \mid \tilde{L} \mid \overline{y}> \;=\; <\overline{y} \mid \overset{\approx}{\tilde{L}} \mid \overline{x}> \;=\; <\overline{y} \mid L \mid \overline{x}> .$$

Consequently,

$$<x \mid L \mid y> \;=\; <\overline{y} \mid \overset{\sqcap}{L} \mid \overline{x}> .$$

And because

$$<x \mid L \mid y> \;=\; <y \mid \tilde{L} \mid x> ,$$

we derive the general transposition rule:

$$\langle x \,|\, \overline{L} \,|\, y \rangle = \langle \bar{x} \,|\, \tilde{L} \,|\, \bar{y} \rangle$$

Contraposition is an important way of applying the implication operator. Because implication is a contrasymmetric operator, invariant to converse transposition, the contraposition law follows directly from the contraposition rule:

$$\langle x \,|\, \rightarrow \,|\, y \rangle = \langle \bar{y} \,|\, \overset{\approx}{\rightarrow} \,|\, \bar{x} \rangle = \langle \bar{y} \,|\, \rightarrow \,|\, \bar{x} \rangle \,.$$

In order to derive the contraposition law we can also explore the transposition rule in combination with the orthogonal transformation rule:

$$\begin{aligned} \langle x \,|\, \rightarrow \,|\, y \rangle &= \langle y \,|\, \overset{\sim}{\rightarrow} \,|\, x \rangle = \langle y \,|\, \leftarrow \,|\, x \rangle = \\ &= \langle \bar{y} \,|\, \theta \leftarrow \,|\, \bar{x} \rangle = \langle \bar{y} \,|\, \rightarrow \,|\, \bar{x} \rangle \,. \end{aligned}$$

The transformation rules in matrix logic take the form of rigorous mathematical derivations. So long as the mathematical rules are observed, there are multiple ways of deriving operator logic equalities.

The concept of the rotation of logic space is instrumental in logic transformations and in the interpretation of inference as computation. Whereas in conventional logic the transformations of logic expressions are heuristic in nature, in matrix logic they are formalized as symmetry transformations in logic space, thus becoming fundamental and rigorously defined.

Matrix logic base

In the previous section we were primarily concerned with multiplicative aspects of matrix logic. However, important features of logic become apparent in the joint approach which explores both the multiplicative and the additive proporties of the formalism.

When considering the additive properties of logic operators, just as was done for forbidden products, we must also introduce the forbidden additions, which implies that the sum of two logic operators must again be a logical operator:

$$L_i + L_j = L_k \,; \qquad L_i, L_j, L_k \in \Omega^+ .$$

The constraint we impose upon additions of logic operators reduces the number of possible sums to a modest set of 50 additions, as compared with 207 possible products. Among these, several types of additive operations are of fundamental importance in matrix logic. Within the strictly multiplicative framework, in order to reduce the major basis set $\{\rceil, \wedge, \vee\}$ to one of the minor sets $\{\rceil, \wedge\}$ or $\{\rceil, \vee\}$, we required the De Morgan rules. Now, with the introduction of the addition operation, these rules cease to be necessary to obtain the required reduction. Since the operators of the major basis set can be linked in an additive way:

$$\rceil + \wedge = \vee ,$$

the disjunction and conjunction operators can be expressed via each other directly:

$$\begin{aligned} \langle x | \vee | y \rangle &= \langle x | \rceil + \wedge | y \rangle = \langle x | \rceil | y \rangle + \langle x | \wedge | y \rangle \\ \langle x | \wedge | y \rangle &= \langle x | \vee - \rceil | y \rangle = \langle x | \vee | y \rangle - \langle x | \rceil | y \rangle . \end{aligned}$$

Consequently, employing matrix addition/subtraction, we can reduce the major basis set to minor sets and vice versa

$$\begin{aligned} \{\rceil, \wedge, \vee\} &\Leftrightarrow \{\rceil, \wedge, \rceil + \wedge\} \\ \{\rceil, \wedge, \vee\} &\Leftrightarrow \{\rceil, \vee - \rceil, \vee\} \end{aligned}$$

What is probably the most significant result that emerges from the analysis of the additive properties of matrix logic is the possibility of a radically new definition of the basis set of logic operators, whereby it is defined in a direct algebraic fashion as an algebraic base, that is, a set of linearly independent matrices from which, with appropriately chosen coefficients, any logic operator can be generated in an additive way. Such a matrix logic base is given by the set of linearly independent operators:

$$\wedge = \begin{pmatrix} 0 & 0 \\ 0 & 1 \end{pmatrix}, \quad \backslash = \begin{pmatrix} 0 & 0 \\ 1 & 0 \end{pmatrix}, \quad / = \begin{pmatrix} 0 & 1 \\ 0 & 0 \end{pmatrix}, \quad \uparrow = \begin{pmatrix} 1 & 0 \\ 0 & 0 \end{pmatrix}.$$

Defining coefficients in the set $E_2 = \{0,1\}$ we can generate the complete set of logic operators by appropriate linear combination of the base operators.

This implies that the matrix logic base $\{\wedge, \backslash, /, \uparrow\}$ is a functionally complete logic basis.

Generators of operators

The matrix logic base allows one to generate the complete set of logic operators. In spite of the fundamental significance of this algebraic matrix approach, there exist even more simple generators of logic. Contrary to the conventional theory, which requires both logic variables and logic connectives as two distinctly different and two necessary ingredients from which a logic calculus can be developed, matrix logic can be generated from logic vectors alone, without connectives or their matrix equivalents being available in advance. All that is needed to generate the logic in all its complexity is a set of two vectors from two adjoint logic vector spaces:

$$|x\rangle \quad \text{and} \quad \langle y| .$$

The inner product of logic vectors $\langle x|y\rangle$ is the Kronecker function

$$\langle x|y\rangle = \delta_{xy}$$

which is a scalar, and which takes us down to the scalar Boolean set $E_2 = \{0,1\}$.

But considering the possible outer products of logic vectors we get :

$$|1\rangle\ \langle 1| = \begin{pmatrix} 0 & 0 \\ 0 & 1 \end{pmatrix}$$

$$|1\rangle\ \langle 0| = \begin{pmatrix} 0 & 0 \\ 1 & 0 \end{pmatrix}$$

$$|0\rangle\ \langle 1| = \begin{pmatrix} 0 & 1 \\ 0 & 0 \end{pmatrix}$$

$$|0\rangle\ \langle 0| = \begin{pmatrix} 1 & 0 \\ 0 & 0 \end{pmatrix}$$

With the help of logic vectors we have generated the matrix set $\{\wedge, \backslash, /, \uparrow\}$, which, as we already know, is the algebraic base of logic, from which any logic operator is derivable as a linear combination with appropriate multipliers. Matrix logic not only transforms conventional connectives into genuine mathematical objects, it also introduces a simple and elegant way to derive them as outer products of the two adjoint vector spaces. With the newly defined generators of logic, the division into logic variables and logic connectives becomes even more artificial. This result has profound consequences for the design of logic and for an understanding of the nature of logic operations. No less important are the implications for computer science. Whereas in conventional computing theory it is common to divide computing elements into operators and operands, the newly defined generators of logic pave the way to a computing theory in which such a division is irrelevant.

Variable logic operators

Throughout our investigation we considered logic functions whose arguments were logic vectors, while logic operators were viewed as constant quantities. In matrix logic a new possibility arises for the reverse formulation, which allows us to enrich the notion of logic operators by considering not only constant but also variable logic operators. Both constant and variable logic operators are generated as the outer products of logic vectors; however, while in the first case the values of the logic vectors are known and constant, in the second case the logic vectors are unknown variables.

Expressing the logic operator via the outer product of two constant logic vectors, we collect logic variable vectors into a variable logic operator and replace variable vector arguments with constant logic vectors extracted from the logic operator.

Consider the case of the conjunction operator

$$\langle x \mid \wedge \mid y\rangle \;=\; \langle x \mid 1\rangle \langle 1 \mid y\rangle \;=\; \langle 1 \mid x\rangle \langle y \mid 1\rangle \;=\; \langle 1 \mid L(\mid x\rangle \langle y \mid) \mid 1\rangle\,,$$

where $L(\mid x\rangle\langle y\mid)$ is a variable logic operator, taking different shapes, depending on the values obtained by the vectors $\mid x\rangle$ and $\langle y \mid$.

In the discrete universe with complementation, a variable logic operator takes values in the 14 operator set:

$$L(\mid x\rangle \langle y \mid) \in \{\wedge, \backslash, /, \uparrow, \perp, \mathrm{T}, \dashv, \vdash, \hat{1}, \hat{0}, \vee, \rightarrow, \leftarrow, \downarrow\}$$

$$\mid x\rangle, \mid y\rangle, \in \{\mid 0\rangle, \mid 1\rangle, \mid \mathbf{0}\rangle, \mid \mathbf{1}\rangle\}$$

The negation operator $\urcorner$ and the identity operator I are excluded from the set, because they cannot be generated by a single outer product.

Vector logic

Employing the new generators of logic, we are able to obtain a pure vector representation of logic functions, in which connective operators do not figure explicitly.

The simplest case of vector representations is found for the equivalence and the negation functions, derived as follows:

$$\langle x \mid \leftrightarrow \mid y\rangle \;=\; \langle x \mid y\rangle$$

$$\langle x \mid \urcorner \mid y\rangle \;=\; \overline{\langle x \mid y\rangle}\;.$$

Replacing the operators by corresponding outer products, we construct the vector representation for logic functions of the matrix logic base:

$$\begin{aligned}
\langle x|\wedge|y\rangle &= \langle x|1\rangle\langle 1|y\rangle \\
\langle x|\backslash|y\rangle &= \langle x|1\rangle\langle 0|y\rangle \\
\langle x|/|y\rangle &= \langle x|0\rangle\langle 1|y\rangle \\
\langle x|\uparrow|y\rangle &= \langle x|0\rangle\langle 0|y\rangle
\end{aligned}$$

In order to derive the vector representation for other functions we employ the complement operator:

$$\begin{aligned}
\langle x|\downarrow|y\rangle &= \langle\overline{x|1\rangle\langle 1|y}\rangle \\
\langle x|\rightarrow|y\rangle &= \langle\overline{x|1\rangle\langle 0|y}\rangle \\
\langle x|\leftarrow|y\rangle &= \langle\overline{x|0\rangle\langle 1|y}\rangle \\
\langle x|\vee|y\rangle &= \langle\overline{x|0\rangle\langle 0|y}\rangle
\end{aligned}$$

Just like the vector representations of the equivalence and negation functions, the vector representations of the filters $\perp$, T, $\dashv$, $\vdash$ are also each reduced to a single inner product:

$$\begin{aligned}
\langle x|\perp|y\rangle &= \langle x|1\rangle \\
\langle x|\mathrm{T}|y\rangle &= \langle x|0\rangle \\
\langle x|\dashv|y\rangle &= \langle 1|y\rangle \\
\langle x|\vdash|y\rangle &= \langle 0|y\rangle
\end{aligned}$$

The vector representation of the empty operator and the tautology operator require the undecided logic vectors:

$$\begin{aligned}
\langle x|\hat{0}|y\rangle &= \langle x|\mathbf{0}\rangle\langle\mathbf{0}|y\rangle \\
\langle x|\hat{1}|y\rangle &= \langle x|\mathbf{1}\rangle\langle\mathbf{1}|y\rangle .
\end{aligned}$$

The octet of logic functions is involutive:

$$\begin{aligned}
\langle x|\leftrightarrow^n|y\rangle &= \langle x|y\rangle^n = \langle x|y\rangle \\
\langle x|\wedge^n|y\rangle &= \langle x|1\rangle\langle 1|1\rangle\langle 1|\ldots|1\rangle\langle 1|y\rangle = \langle x|1\rangle\langle 1|y\rangle \\
\langle x|\uparrow^n|y\rangle &= \langle x|0\rangle\langle 0|0\rangle\langle 0|\ldots|0\rangle\langle 0|y\rangle = \langle x|0\rangle\langle 0|y\rangle \\
\langle x|\perp^n|y\rangle &= \langle x|1\rangle^n = \langle x|1\rangle \\
\langle x|\mathrm{T}^n|y\rangle &= \langle x|0\rangle^n = \langle x|0\rangle \\
\langle x|\dashv^n|y\rangle &= \langle 1|y\rangle^n = \langle 1|y\rangle \\
\langle x|\vdash^n|y\rangle &= \langle 0|y\rangle^n = \langle 0|y\rangle \\
\langle x|\hat{0}^n|y\rangle &= \langle x|\mathbf{0}\rangle\langle\mathbf{0}|\mathbf{0}\rangle\langle\mathbf{0}|\ldots|\mathbf{0}\rangle\langle\mathbf{0}|y\rangle = \langle x|\mathbf{0}\rangle\langle\mathbf{0}|y\rangle ,
\end{aligned}$$

where n is any positive integer.

Vector logic is characterized by a number of remarkable properties which make inference and valuation procedures in matrix logic quite simple, elegant and effective.

The derivation of logic equivalences over the vector logic representation becomes especially easy and transparent. Making use of the identity decomposition rule and the complementation rule of matrix logic, we derive the De Morgan relations as follows:

$$\langle\overline{x\,|\wedge|\,y}\rangle = \langle\overline{x\,|\,1\rangle\langle 1\,|\,y}\rangle = \langle\overline{\bar{x}\,|\,0\rangle\langle 0\,|\,\bar{y}}\rangle = \langle\bar{x}\,|\vee|\,\bar{y}\rangle$$
$$\langle\overline{x\,|\vee|\,y}\rangle = \langle x\,|\,0\rangle\langle 0\,|\,y\rangle = \langle\bar{x}\,|\,1\rangle\langle 1\,|\,\bar{y}\rangle = \langle\bar{x}\,|\wedge|\,\bar{y}\rangle$$

Applying the two rules in one order for the first equation, and in reverse order for the second equation, we obtain the matrix logic formulation of the De Morgan laws.

Clearly, the principles established here can be used for the derivation of a wide range of logic equivalences, with the possibility of very economical computation.

2.2.2 Matrix strings

Up to this point we have dealt with the basic logic functions. Our next task is to investigate how matrix logic can be applied to the computation of compound logic expressions or strings. In conventional logic the main method of evaluating large logic strings is based on reducing them to disjunctive or conjunctive normal forms and then minimizing them. In developing methods for the computation of large logic strings in matrix logic we face not only new opportunities, provided by the richness and power of matrix calculus, but also several major obstacles ascribable to the limitations imposed by the rules of matrix multiplication. Within the framework of the matrix formulation, a logic string represents the product of a finite number of matrices, that is logic operators and logic vectors, which, to be multiplied, must obey the rules of matrix multiplication. To fulfil the condition of matrix multiplicaton it is necessary that each pair of neighbouring matrices on a string should be compatible, that is, in each succeeding matrix the number of rows must be equal to the number of columns in the preceding matrix. It is not difficult to write down a compound logic expression the corresponding matrix form of which would not comply with the essential condition for matrix multiplication. Nevertheless, the matrix formulation of logic allows us to construct fundamental representations for compound logic functions of any complexity.

Conjunction operator

Boolean logic algebra is designed over the operations of conjunction $\wedge$ and disjunction $\vee$ which are put in correspondence with the ordinary operations of multiplication and addition. The analysis we have carried out in the preceding sections shows clearly that matrix logic represents more complex relationships between numerical and logical operations. The domain of matrix logic is incomparably wider and is not restricted to the operations $\wedge$ and $\vee$, but includes the entire set of logic operators.

Although matrix logic algebra is designed similarly over operations of matrix multiplication and addition, these matrix operations are *superimposed* over logical operations. It is important to recognize that whereas in Boolean algebra $\wedge$ and $\vee$ represent operations, in matrix logic algebra these are matrix operators, and there are no particular reasons for their special treatment with respect to other logic operators.

Nevertheless, the role of the conjunction operator in matrix logix is both unique and fundamental. It is this role and the special properties of conjunction that allow us to obtain a canonical reduction for compound matrix logic expressions.

Because Boolean logic algebra identifies conjunction with the physically correct operation of the multiplication of binary numbers—

$$
\begin{aligned}
0 \cdot 0 &= 0 \\
1 \cdot 0 &= 0 \\
0 \cdot 1 &= 0 \\
1 \cdot 1 &= 1
\end{aligned}
$$

—this operation is naturally retained in matrix logic whenever conjunction of scalar logic quantities is required. At the same time we also employ conjunction as a matrix operator whenever logic quantities are in vector form. Consequently, conjunction can appear in matrix logic in two different mathematical forms: as the operation of number multiplication and as a matrix operator. In addition to making use of the operator form of conjunction, matrix logic also preserves its Boolean quality as involving the multiplication of numbers. This not only bridges in an interesting way Boolean logic and matrix logic, but more importantly the dual nature of conjunction discussed above is instrumental in the reduction of compound logic expressions or matrix logic strings to the canonical form of the fundamental product.

Fundamental product

A fundamental product is a product of atomic bracket products, in the general case an algebraic sum:

$$\sum^{n} \bigwedge^{m} < x_i \,|\, L_j \,|\, y_k > ,$$

where $<x_i|$ are variable or constant vectors from the bra logic vector space, $|y_k>$ are variable or constant vectors from the ket logic vector space, and L_j are logic operators.
The fundamental product is a representation that has profound significance, implying that logic expressions can be reduced to the standard uniform form of an algebraic sum of products of the atomic valuation devices of matrix logic. The possibility of joining the basic atomic products into a fundamental product is a direct result of the special dual properties of conjunction. Whenever two atomic products in a logic expression are not linked by a conjunction, we transform the expression to obtain the conjunction $\wedge$ and then delete the operator, since multiplication is denoted simply by the juxtaposition of atomic products.

Kronecker logic product

By using the outer product representation for logic operators we are able to achieve the reduction of a fundamental product to a corresponding representation which can be defined as the Kronecker logic product.
The Kronecker logic product is an algebraic sum of matrix strings, each representing a product of Kronecker delta functions:

$$< x_1 \,|\, y_1 >< x_2 \,|\, y_2 >< x_3 \,|\, y_3 > \ldots < x_n \,|\, y_n > \;=\; \delta_{x_1 y_1}\, \delta_{x_2 y_2}\, \delta_{x_3 y_3} \ldots \delta_{x_n y_n} ,$$

where $<x_i|$ and $|y_i>$ are variable or constant logic vectors.
In the Kronecker delta reduction of logic expressions it becomes apparent that scalar logic values are actually scalar tensors — a fact which conventional scalar logic would not be able to either detect or explain.

Operator strings

The unique features of matrix operator logic and its fundamentally new ways of evaluating logic expressions are demonstrated by the operator string problem. It is here that the exceptional computational power of matrix logic shows itself most forcefully both from the theoretical point of view and from the point of view of applications. The connective operators in matrix logic strings can interact directly, a property inconceivable for conventional logic computations. Let us consider a particular fundamental product comprising disjunction and implication brackets:

$$\langle x \mid \vee \mid 1\rangle \langle 1 \mid \rightarrow \mid y\rangle = 1 \cdot y = y \,.$$

Normally, we evaluate each atomic product and multiply the results obtained according to the conjunction rule. But we could equally well have proceeded differently:

$$\langle x \mid \vee \mid 1\rangle \langle 1 \mid \rightarrow \mid y\rangle = \langle x \mid \vee \wedge \rightarrow \mid y\rangle = \langle x \mid \dashv \mid y\rangle = y \,.$$

Here we have transformed the expression into an operator string and solved the problem by calculating the product of operators. This procedure is what we defined as the operator string problem, which generally involves finding a solution to strings of the following structure:

$$\langle x \mid L_1 \, L_2 \, L_3 \ldots L_n \mid y\rangle \,,$$

where $L_1 \, L_2 \, L_3 \ldots L_n$ are constant or variable logic operators, in the first case producing one of the basic logic operators, in the second case generating a compound logic operator.
The resolution of the operator string problem does not involve any particular complications. It has a direct solution: carrying out multiplication of the operators $L_1, L_2, L_3, \ldots L_n$, we obtain the resulting operator and can compute the value of the logic function in question.
Operator strings are yet another important canonical reduction of the fundamental product. Considering an operator string as a logic hyperfunction, we obtain a converse hyperfunction and a contraposed hyperfunction by applying the standard matrix logic operations:

converse string: $\langle x \mid \widetilde{L_1 \, L_2} \mid y \rangle = \langle x \mid \tilde{L}_2 \, \tilde{L}_1 \mid y \rangle$

contraposition string: $\langle x \mid \overline{L_1 \, L_2} \mid y \rangle = \langle x \mid \overline{L_2} \, \overline{L_1} \mid y \rangle$

where transposition and converse transposition of an operator product are a product of transposes or converse transposes taken in reverse order.

Transformation rules

The transformation rules for matrix logic strings can be divided into two major classes: deductive and inductive transformation rules.

1. Superassociativity

The associativity of multiplication and the commutativity of the inner products in matrix logic strings lead to a sort of logic associativity: a logic vector can be bound via an operator either to a bra logic vector on the left or to a ket logic vector on the right, with both cases yielding the identical result for the evaluation of an expression:

$$\delta_{x_1y_1}\,\delta_{x_2y_2}\,\delta_{x_3y_3} = \delta_{x_1y_1}\,\tilde{\delta}_{x_2y_2}\,\delta_{x_3y_3} = \delta_{x_1y_1}\,\delta_{y_2x_2}\,\delta_{x_3y_3}$$

Consider a particular logic associativity transformation:

$$\begin{aligned}\langle x | \backslash | y\rangle \langle 1 | z\rangle &= \langle x | 1\rangle \langle 0 | y\rangle \langle 1 | z\rangle = \\ &= \langle x | 1\rangle \langle y | 0\rangle \langle 1 | z\rangle = \\ &= \langle x | 1\rangle \langle y | / | z\rangle .\end{aligned}$$

Furthermore, canonical matrix strings are characterized not only by commutativity and associativity but also by what may be called superassociativity. An inner product particle is freely transferable within a logic expression, unless it is prevented by the complement operator from entering or leaving a particular area of a matrix string. Within this constraint superassociativity allows a given inner product or its copy to be detached from a given place on the matrix string and to be attached in a new place, which induces a new interaction. Superassociativity offers the possibility of deriving new logic relations by the realignment of inner or atomic products in one of the canonical reductions. Due to the associativity of number multiplication the inner product and the atomic product can be detached from their position in a matrix string and attached to a new position. Such a realignment could lead to the formation of new logic operators and to the migration of a logic vector from one atomic product to another. Superassociativity, together with the commutativity of inner products, constitutes in matrix logic an inference procedure of major importance.

Reducing a given logic expression to a Kronecker delta product, and applying the commutativity, associativity and superassociativity transformations, we can recompose the expression in a great variety of equivalent forms. Consider a particular application of superassociativity and commutativity rules:

$$\begin{aligned}\langle x | \wedge | y\rangle \langle z | \uparrow | t\rangle &= \langle x | 1\rangle \langle 1 | y\rangle \langle z | 0\rangle \langle 0 | t\rangle = \\ &= \langle z | 0\rangle \langle 1 | x\rangle \langle y | 1\rangle \langle 0 | t\rangle = \\ &= \langle z | / | x\rangle \langle y | \backslash | t\rangle .\end{aligned}$$

Exploring the superassociativity, logic associativity and commutativity rules, we rearranged the connective operators between logic vectors and plugged in operators between previously separate vectors, without any impact on the value of the logic expression.

2. Idempotence

In the discrete universe due to the idempotence the value of a matrix string is unaltered if the latter is expanded by any number of identical bracket or inner products. Conversely, if an expression contains identical products, which are all either under or outside the complement operator, they all, except for one, can be deleted with no effect on the result of evaluation. Consider the following reductions:

$$<x\,|\wedge|\,x><x\,|\wedge|\,x> \;=\; <x\,|\wedge|\,x> \;=\; <x\,|\,1><1\,|\,x> \;=\; <x\,|\,1>$$

$$<x\,|\vee|\,x><x\,|\vee|\,x> \;=\; <x\,|\vee|\,x> \;=\; \overline{<\bar{x}\,|\,1><1\,|\,\bar{x}>} \;=\; <x\,|\,1>$$

3. Transmultiplication

The transmultiplication rule is the direct application of the transmultiplication operation. It can be employed either for the fusion of operators:

$$<x\,|\,L_1\,|\,y><x\,|\,L_2\,|\,y> \;=\; <x\,|\,L_1 \circledast L_2\,|\,y> \;=\; <x\,|\,L_3\,|\,y>$$

or for the separation of operators:

$$<x\,|\,L_3\,|\,y> \;=\; <x\,|\,L_1 \circledast L_2\,|\,y> \;=\; <x\,|\,L_1\,|\,y><x\,|\,L_2\,|\,y>$$

In the first case several atomic products are fused into a single atomic product. In the second case an atomic product decays into several atomic products. The most important transmultiplication transformation concerns the equivalence function:

$$<x\,|\rightarrow|\,y><x\,|\leftarrow|\,y> \;=\; <x\,|\leftrightarrow|\,y> \;=\; <x\,|\,y> \;=\; \delta_{xy}$$

4. Subtraction of contradiction

Subtraction of contradiction represents the fundamental induction transformation rule. Whereas for deduction rules new logic variables do not appear in transformation, the subtraction-of-contradiction rule allows us to plug, if required, new variables into an expression. Though this induction may not change the logic function realized by a matrix string, the introduction of a new variable followed by a transformation of a string may reveal new properties of an expression which were not evident before.

The general forms of the inductive transformation are as follows:

$$\alpha = \alpha - 0$$

or

$$\alpha = \alpha + 0$$

where α is an arbitrary logic expression, and 0 stands for an arbitrary contradiction. The value of the expression α naturally does not change with the subtraction or addition of null. However, if the contradiction is constructed in such a fashion as to include some new variable or variables, a further transformation of the string could lead to a non-trivial expression that is qualitatively different, though the logic function realized by the string is in no way altered.

The two fundamental contradiction terms which we shall employ further are as follows:

$$<x\,|\,1><y\,|\,1><x\,|\,0> \;\equiv\; 0$$
$$<x\,|\,1><y\,|\,0><x\,|\,0> \;\equiv\; 0$$

where y is a new logic variable.

Implication algebra

An implication algebra provides a complete algebraic characterization of the implication connective, which is fundamental for logic deduction. An implication operator defines an implication algebra, satisfying the transitivity law,

$$(x \to y) \wedge (y \to z) \to (x \to z)$$

and the following conditions:

1. $x \to (y \to z) = y \to (x \to z)$
2. $(x \to y) \to y = (y \to x) \to x$
3. $(x \to y) \to x = x$

First we establish that the fundamental transitivity relation is a tautology:

$$<x\,|{\to}|\,y><y\,|{\to}|\,z> \to <x\,|{\to}|\,z> =$$

$$= \overline{<x\,|{\to}|\,y><y\,|{\to}|\,z>}<x\,|\backslash|\,z> =$$

$$= < \overline{\overline{x | \backslash | y} >< \overline{y | \backslash | z} >< x | \backslash | z} > =$$

$$= < \overline{x | \backslash | z > - < x | \backslash | y >< x | \backslash | z > - < y | \backslash | z >< x | \backslash | z > +}$$
$$\overline{+ < x | \backslash | y >< y | \backslash | z >< x \ \backslash \ z} > =$$

$$= < \overline{x | \backslash | z > - < x | \backslash | y >< x | \backslash | z > - < y | \backslash | z >< x | \backslash | z > + 0} =$$

$$= \overline{< x | \backslash | z > - < x | 1 >< 0 | y >< x | 1 >< 0 | z > - < y | 1 >< 0 | z >< x | 1 >< 0 | z >} =$$

$$= \overline{< x | \backslash | z > - < x | 1 >< 0 | z >< y | 0 > - < x | 1 >< 0 | z >< y | 1 >} =$$

$$= \overline{< x | \backslash | z > \ (1 - < y | 0 > - < y | 1 >)} =$$

$$= \overline{< x | \backslash | z > \cdot 0} = \bar{0} \equiv 1 \qquad \blacksquare$$

The equalities of the implication algebra can be proved either deductively or inductively from the tautology X = X.
We shall employ both deductive and inductive procedures for derivation.

1. <u>Permutation equality $x \rightarrow (y \rightarrow z) = y \rightarrow (x \rightarrow z)$</u>:

By matricization of the symbolic logic expression we obtain:

$$< \overline{x | 1 >< \overline{y | \rightarrow | z}} > = < \overline{y | 1 >< \overline{x | \rightarrow | z}} >$$

and

$$< x | 1 >< \overline{y | \rightarrow | z} > = < y | 1 >< \overline{x | \rightarrow | z} >$$

Eliminating the complement operators,

$$< x | 1 >< y | \backslash | z > = < y | 1 >< x | \backslash | z > \ ,$$

and employing outer product representations of the Inhibition operator, we get

$$< x | 1 >< y | 1 >< 0 | z > = < y | 1 >< x | 1 >< 0 | z > \qquad \blacksquare$$

The permutation relation is proven.
Quite naturally we could have proceeded inductively:

$$\alpha = \alpha$$

$$< x \mid 1 >< y \mid 1 >< 0 \mid z > = < x \mid 1 >< y \mid 1 >< 0 \mid z >$$

Due to commutation of inner products:

$$< x \mid 1 >< y \mid 1 >< 0 \mid z > = < y \mid 1 >< x \mid 1 >< 0 \mid z >$$

Computing the outer products:

$$< x \mid 1 >< y \mid \backslash \mid z > = < y \mid 1 >< x \mid \backslash \mid z >$$

Extracting the complement operators:

$$< x \mid 1 >< \overline{y \mid \rightarrow \mid z} > \; = \; < y \mid 1 >< \overline{x \mid \rightarrow \mid z} >$$

and complementing both sides of the equation we derive:

$$< \overline{x \mid 1 >< \overline{y \mid \rightarrow \mid z}} > \; = \; < \overline{y \mid 1 >< \overline{x \mid \rightarrow \mid z}} >$$

Finally, by scalarization we pass to the conventional symbolic expression:

$$x \rightarrow (y \rightarrow z) \; = \; y \rightarrow (x \rightarrow z) \qquad \blacksquare$$

2. <u>Equality $(x \rightarrow y) \rightarrow y \; = \; (y \rightarrow x) \rightarrow x$</u>:

Matrix logic mapping:

$$< \overline{x \mid \rightarrow \mid y >< \overline{1 \mid y}} > \; = \; < \overline{y \mid \rightarrow \mid x >< \overline{1 \mid x}} >$$

and

$$< x \mid \rightarrow \mid y >< 0 \mid y > \; = \; < y \mid \rightarrow \mid x >< 0 \mid x >$$

Outer product representation of the implication operator:

$$< \overline{x \mid 1 >< 0 \mid y} >< 0 \mid y > \; = \; < \overline{y \mid 1 >< 0 \mid x} >< 0 \mid x >$$

Elimination of the complement operators:

$$<y \mid 0> - <x \mid 1><y \mid 0>^2 \; = \; <x \mid 0> - <y \mid 1><x \mid 0>^2$$

Due to idempotence:

$$<y \mid 0> - <x \mid 1><y \mid 0> \; = \; <x \mid 0> - <y \mid 1><x \mid 0>$$

Expulsion of common multipliers:

$$<y \mid 0>(1 - <x \mid 1>) \; = \; <x \mid 0>(1 - <y \mid 1>) \, ,$$

that is,

$$<y \mid 0><x \mid 0> \; = \; <x \mid 0><y \mid 0> \, . \qquad \blacksquare$$

The equality is proven.

3. Equality $(x \rightarrow y) \rightarrow x = x$:

An inductive proof of this equality employs the important procedure of subtraction of the null term, that is, a contradiction which introduces a new variable.
In the general tautology form,

$$\alpha = \alpha$$

we substitute

$$<x \mid 0> \; = \; <x \mid 0> \, ,$$

which is clearly equivalent to

$$<x \mid 0> - 0 \; = \; <x \mid 0> \, ,$$

since addition or subtraction of a contradiction is algebraically irrelevant. But we now construct a null term in such a way that it contains a new variable y:

$$\langle x \mid 0\rangle - \langle x \mid 1\rangle\langle 0 \mid y\rangle\langle x \mid 0\rangle = \langle x \mid 0\rangle$$

Expelling the common multiplier $\langle x \mid 0\rangle$, we have

$$(1 - \langle x \mid 1\rangle\langle 0 \mid y\rangle)\,\langle x \mid 0\rangle = \langle x \mid 0\rangle$$

$$\overline{\langle x \mid 1\rangle\langle 0 \mid y\rangle}\langle x \mid 0\rangle = \langle x \mid 0\rangle$$

Elimination of the complement yields:

$$\langle x \mid \rightarrow \mid y\rangle\langle x \mid 0\rangle = \langle x \mid 0\rangle$$

And

$$\langle x \mid \rightarrow \mid y\rangle\overline{\langle x \mid 1\rangle} = \langle x \mid 0\rangle$$

Complementing both sides of the equation:

$$\overline{\langle x \mid \rightarrow \mid y\rangle\overline{\langle x \mid 1\rangle}} = \overline{\langle x \mid 0\rangle}$$

$$\overline{\langle x \mid \rightarrow \mid y\rangle\overline{\langle x \mid 1\rangle}} = \langle x \mid 1\rangle$$

and by scalarization we obtain

$$(x \rightarrow y) \rightarrow x = x \qquad \blacksquare$$

2.2.3 Intelligence code

The inclusion of logic theory in a general system of covariant laws is particularly important in a framework of the study of high-level intelligence. Clearly, if cognition is adequate to physical reality, then its laws must satisfy the general covariance relations. From this point of view the intelligence code is defined as a system of logic operations which are invariant under arbitrary rotations of logic space, realized by physical interactions.

The most general view assumes that the intelligence code is fundamental and universal. Though two intelligent systems may be based on different rules of inference, because of the covariance property both shall derive, say, the Pythagoras theorem, though each may employ its own language and its own mental dictionary.

The idea that the intelligence code results from an elaborate knowledge and language capability is a major misconception. A logic system may possess most valuable information and data, yet it may be unable to display intelligent functions. The data and language alone do not yet license the power of intelligent operations.

What distinguishes the intelligent mind from ordinary data or information processing is the availability of rules of inference. This includes both deductive and inductive rules, which are typical of high-level intelligence. Because all rules of inference are necessarily tautologies, the intelligence code can be described as the covariant system of rules of inference, which is in itself a global tautology.

Tautologies are not only a universal code of communication, only tautologies are covariant and only true results of intelligence are tautologies. It would be no exaggeration to say that the power of the human mind results in one way or another from the remarkable ability to obtain and retain tautologies. Thus, in logic terms, the problem of the covariance of high-level intelligence can be reformulated as the problem of the formation and transformation of tautologies. Because much of the intelligence code puzzle can be traced to the fundamental logic construct of tautology, the study of the properties of tautologies, as well as the rules that lead to their formation, is instrumental in the deciphering of the intelligence code.

Considering tautologies as elements and coding units of the intelligence code, we can formulate the underlying algorithms for constructing these coding units. In the general logic framework the units of the intelligence code are theorems, since every valid theorem of logic theory is a tautology and vice versa.

The description of the intelligence code as an invariant system of tautologies does not yet make it a subject of mathematical theory. This is achieved after the expressions are associated with exact numerical measures in the logic space.

Vector and operator tautologies

In matrix logic we distinguish two different classes of tautology: vector tautologies and operator tautologies. The vector tautology is a logic expression which remains invariant under different values taken by arguments. This is the case with the following atomic bracket products:

$$\forall x: \quad \langle x | \leftrightarrow | x \rangle \equiv 1$$
$$\langle x | \vee | \bar{x} \rangle \equiv 1$$
$$\langle x | \rightarrow | x \rangle \equiv 1$$

The operator tautology is a logic expression which is invariant under the substitution of different logic operators. Consider

$$\langle x | L | x \rangle + \langle x | \bar{L} | x \rangle \equiv 1 .$$

Clearly, the expression is a tautology, satisfied by any operator:

$$< x \mid L \mid y > = 1 - < x \mid \bar{L} \mid y >$$

$$< x \mid L \mid y > = \overline{< x \mid \bar{L} \mid y >}$$

$$< x \mid L \mid y > = < x \mid L \mid y >$$

<u>Negation of contradiction</u>

Another fundamental property of the intelligence code is reflected in the existence of two major subclasses of vector tautologies. The first subclass can be defined as the direct or saturated tautologies. This is the case for the tautology operator

$$<x \mid \hat{1} \mid y> \equiv 1 ,$$

and for the inner product of a logic vector

$$<x \mid x> \equiv 1 ,$$

which is in fact the equivalence logic function applied to a single variable:

$$<x \mid \leftrightarrow \mid x> = <x \mid x> \equiv 1 .$$

The second and much larger subclass we define as anti-contradiction tautologies. Because a tautology is derivable through negation of the associated contradiction, the appearance of a contradiction automatically paves the way for the construction of a tautology. Here we detect an important procedure: the intellect grows by discovering and resolving contradictions. Earlier we established that the analysis of tautologies and the study of the intelligence code are directly related; in much the same way we can now descend to the level of contradictions, since their analysis is similarly useful for the formulation of the intelligence code.
At the level of vector logic the concept of contradiction is a consequence of orthogonality: whenever logic vectors are orthogonal, the corresponding inner vector product is a contradiction:

$$<x \mid \bar{x}> \equiv 0 ,$$

and its negation is necessarily a tautology:

$$\overline{< x \mid \bar{x} >} \equiv 1 .$$

Multiplicative and additive contradictions

In matrix logic, contradictions can be formed in a multiplicative or additive manner. The additive form of a contradiction or tautology we derive from the contradiction law and the law of the excluded middle respectively:

$$\overline{\alpha\,\bar{\alpha}} = \begin{cases} \overline{\alpha\,(1-\alpha)} = \overline{\alpha-\alpha^2} = \overline{\alpha-\alpha} = \bar{0} \equiv 1 \\ \overline{\bar{\bar{\alpha}}\,\bar{\alpha}} = \overline{1-\bar{\alpha}-\alpha+\underbrace{\bar{\alpha}\,\alpha}_{0}} = \overline{1-\bar{\alpha}-\alpha} \equiv 1 \end{cases}$$

An important case of multiplicative contradiction is a scaled form of the contradiction law, $\alpha\,\bar{\alpha} \equiv 0$,

$$\forall\beta: \qquad \beta\,\alpha\,\bar{\alpha} \equiv 0\,.$$

The negation of particular reductions of this form lead to the two fundamental inference rules:

Anything from false:

$$\bar{x} \rightarrow (x \rightarrow y) \equiv 1$$

True from anything:

$$x \rightarrow (y \rightarrow x) \equiv 1$$

The multiplicative forms of the tautology and contradiction are tremendously effective because by employing multiplicative representations we may not only avoid exponential explosions, but also achieve a drastic reduction of computing for a certain subset of problems. The detection of a single zero multiplier in an expression of any complexity automatically determines the result of the evaluation and renders further calculation irrelevant and unnecessary. In evaluating the validity of logic products this leads to a vast simplification.

Implication tautology

The mind possesses a fluent and elaborate capacity for designing new tautologies. The equivalence transformation can be employed to derive the important class of implication tautologies. Given a logic expression α, we translate it into an equivalent logic expression β:

$$\alpha = \beta\,.$$

Then

$$\alpha \to \beta$$

is a tautology,

$$\overline{\alpha\overline{\overline{\beta}}} \equiv 1 ,$$

which can be employed as a new inference rule.
On the other hand, given an expression $\overline{\alpha\overline{\overline{\beta}}}$, we must verify whether α equals β, and if so $\overline{\alpha\overline{\overline{\beta}}}$ is a tautology. Indeed, from

$$(x = y) \to (x \to y)$$

we get by matricization

$$\langle\overline{x|y\rangle\langle\overline{x|\to|y}}\rangle = \langle\overline{x|y\rangle\langle x|\backslash|y}\rangle .$$

Employing transmultiplication fusion of the operators, we verify that the expression is a tautology:

$$\begin{aligned} \langle\overline{x|y\rangle\langle x|\backslash|y}\rangle &= \langle\overline{x|\leftrightarrow \ast \backslash|y}\rangle = \\ &= \langle\overline{x|\hat{0}|y}\rangle = \\ &= \langle x|\hat{1}|y\rangle \equiv 1 \qquad \blacksquare \end{aligned}$$

<u>Transmultiplication tautology</u>

In comparison to designing new tautologies serving as coding units of the intelligence code, the problem of verifying whether a given logic construct is a tautology represents a much more difficult problem. One of the effective tools which can be employed for this verification is the transmultiplication operation.
Consider the classical rule of inference through a reduction to the absurd: if $\overline{x}$ implies both y and $\overline{y}$, then $\overline{\overline{x}}$ is true:

$$[(\bar{x} \to y) \wedge (\bar{x} \to \bar{y})] \to x$$

$$\langle \overline{\bar{x} \mid \to \mid y \rangle \langle \bar{x} \mid \to \mid \bar{y} \rangle \langle \overline{x \mid 1}} \rangle \;=$$

$$= \langle \overline{x \mid \vee \mid y \rangle \langle x \mid \leftarrow \mid y \rangle \langle x \mid 0} \rangle \;=$$

$$= \langle \overline{x \mid \vee \ast \leftarrow \mid y \rangle \langle x \mid 0} \rangle \;=$$

$$= \langle \overline{x \mid \perp \mid y \rangle \langle x \mid 0} \rangle \;=$$

$$= \langle \overline{x \mid 1 \rangle \langle x \mid 0} \rangle \;= \bar{0} \equiv 1$$

By merging the implications in the initial expression, we avoid the need for transmultiplication fusion. The expression is reduced to the canonical form as follows:

$$(\bar{x} \to (y \wedge \bar{y})) \to x$$

$$\langle \overline{\overline{\bar{x} \mid 1 \rangle \langle \overline{y \mid \wedge \mid \bar{y}}} \rangle \langle \overline{1 \mid x}} \rangle \;=$$

$$= \langle \overline{\overline{x \mid 0 \rangle \langle y \mid \downarrow \mid \bar{y}} \rangle \langle 0 \mid x} \rangle \;=$$

$$= \langle \overline{\overline{x \mid 0 \rangle \langle y \mid \to \mid y} \rangle \langle 0 \mid x} \rangle \;=$$

$$= \langle \overline{x \mid 1 \rangle \langle 0 \mid x} \rangle \;= \bar{0} \equiv 1$$

The structure of tautology may not be immediately recognizable in the expression of two logic variables:

$$\langle \overline{x \mid \vee \mid \bar{y} \rangle \langle \bar{x} \mid \vee \mid y \rangle \langle x \mid \vee \mid y \rangle \langle \bar{x} \mid \vee \mid \bar{y}} \rangle \;.$$

However, using the transmultiplication rule, we can reduce this expression to an explicit canonical form:

$$\langle \overline{x \mid \vee \mid \bar{y} \rangle \langle \bar{x} \mid \vee \mid y \rangle \langle x \mid \vee \mid y \rangle \langle \bar{x} \mid \vee \mid \bar{y}} \rangle \;=$$

$$= \langle \overline{x \mid \leftarrow \mid y \rangle \langle x \mid \to \mid y \rangle \langle x \mid \vee \mid y \rangle \langle x \mid \downarrow \mid y} \rangle \;=$$

$$= \langle \overline{x \mid \leftrightarrow \mid y \rangle \langle x \mid \rceil \mid y} \rangle \;=$$

$$= \langle \overline{x \mid y \rangle \langle \overline{x \mid y}} \rangle \;\equiv 1$$

The derived tautology belongs to the class of tautologies with two logic variables of the general form

$$\langle \overline{x \mid L \mid y \rangle \langle x \mid \bar{L} \mid y} \rangle \;,$$

where L and $\bar{L}$ are two complementary logic operators.

A transmultiplication of an operator and its complement is automatically the empty operator. However, the empty operator can result in transmultiplication fusion of operators that are not necessarily complements of each other. Consider

$$\langle x | \mathrm{T} | y \rangle \langle x | \wedge | y \rangle = \langle x | \mathrm{T} \circledast \wedge | y \rangle = $$
$$= \langle x | \hat{0} | y \rangle \equiv 0$$

The contradiction here can also be established by reduction to the fundamental annihilating term because the value of the antifilter bracket is indifferent to the value of y:

$$\langle x | \mathrm{T} | y \rangle \langle x | \wedge | y \rangle = \langle x | 0 \rangle \langle x | 1 \rangle \langle 1 | y \rangle \equiv 0$$

The presence of the filters in tautology or contradiction expressions opens the possibility of the inclusion or exclusion of new variables, thus providing a basis for inductive logic operations.

Matrix base tautologies

Substituting the operators of the matrix logic base $\{ \wedge, \backslash, /, \uparrow \}$ into the canonical form, we obtain four tautology expressions which play a fundamental role in the operations of the intelligence code:

(1) $\langle \overline{x | \backslash | y \rangle \langle x | \rightarrow | y} \rangle \equiv 1$

(2) $\langle \overline{x | \uparrow | y \rangle \langle x | \vee | y} \rangle \equiv 1$

(3) $\langle \overline{x | / | y \rangle \langle x | \leftarrow | y} \rangle \equiv 1$

(4) $\langle \overline{x | \wedge | y \rangle \langle x | \downarrow | y} \rangle \equiv 1$

By scalarization and by making use of the superassociativity and commutativity rules, we can obtain the classical inference rules:

(1) Modus Ponens rule

$(x \wedge (x \rightarrow y)) \rightarrow y$

$$\langle \overline{x | 1 \rangle \langle x | \rightarrow | y \rangle \langle 0 | y} \rangle = \langle \overline{x | 1 \rangle \langle 0 | y \rangle \langle x | \rightarrow | y} \rangle =$$
$$= \langle \overline{x | \backslash | y \rangle \langle x | \rightarrow | y} \rangle \equiv 1 .$$

Semantic reading: *x is true and x implies y, then y is true.*

(2) Modus Tollens rule

$((\overline{x} \to y) \wedge \overline{y}) \to x$

$$<\overline{\overline{x} \mid \to \mid y><1 \mid \overline{y}><0 \mid x}> \; = \; <\overline{x \mid 0><0 \mid y><x \mid \vee \mid y}> \; =$$

$$= \; <\overline{x \mid \uparrow \mid y><x \mid \vee \mid y}> \; \equiv 1 \, .$$

Semantic reading: $\overline{x}$ *implies y and y is false, then x is true.*

(3) Converse Modus Ponens rule

$((x \leftarrow y) \wedge y) \to x$

$$<\overline{x \mid \leftarrow \mid y><1 \mid y><0 \mid x}> \; = \; <\overline{x \mid 0><1 \mid y><x \mid \leftarrow \mid y}> \; =$$

$$= \; <\overline{x \mid / \mid y><x \mid \leftarrow \mid y}> \; \equiv 1 \, .$$

Semantic reading: *x if y and y is true, then x is true.*

(4) Converse Modus Tollens rule

$(x \wedge (y \to \overline{x})) \to \overline{y}$

$$<\overline{x \mid 1><y \mid \to \mid \overline{x}><1 \mid y}> \; = \; <\overline{x \mid 1><1 \mid y><x \mid \downarrow \mid y}> \; =$$

$$= \; <\overline{x \mid \wedge \mid y><x \mid \downarrow \mid y}> \; \equiv 1 \, .$$

Semantic reading: *x is true and y implies x false,* $\overline{x}$*, then y is false,* $\overline{y}$.

So far we have been primarily concerned with mapping logic expressions onto mathematical structures. The inverse mapping is of no less significance. The extreme importance of the intelligence code resides in the fact that the tautological structures of the code allow for a direct and universally valid content translation. Because tautologies represent fundamental invariants of cognition, the problem of the intelligence code is put in direct relation to the question of logical truth. Independently of a particular interpretation given to the coding units of the code its properties remain invariant. Consequently, the overriding issue for the theory of the intelligence code is the conservation of logic values. From this perspective the intelligence code can be effectively defined as a fundamental law of conservation of logical truth (information), with a status analogous to the laws of conservation of energy or momentum. Whereas the latter result from the time and space isotropy, the former is due to the isotropy of information as realized by the logic space. In this a distinction is made between that which is true or false by assignment, and that which is true or false by being actually realized. At the end of this chapter we shall be in a position to transform these general considerations into rigorous mathematics.

2.3 Continuous universe $\Omega^{\Diamond}$

2.3.1 Semitruth

Any study of logic would be grossly incomplete if it failed to account for the indeterministic mode of reasoning. The introduction of matrix operators in place of logic connectives provides constructive mathematical means for extensions of logic which otherwise could be conceived only in a postulatory manner, if at all. In a natural way the matrix logic framework supports an important extension of logic into the domain of fractional and continuous logic values. Continuous logic is an extension which deals with approximate or imprecise forms of reasoning, which involve partially true and partially false truth-values. When logic premisses are accepted with incomplete certainty of their truth, they cannot as such provide an absolute guarantee of the truth of the conclusions. Conventional logic attempts to deal with this problem simply by borrowing methods from probability theory. The conventional approach consequently offers no new understanding of indeterministic modal logic, nor does it enhance its computational capabilities significantly.

Since the numerical measures of true and false in a discrete universe are given by the basis vectors, which are binary logic vectors

$$\begin{pmatrix} 0 \\ 1 \end{pmatrix} \quad \text{and} \quad \begin{pmatrix} 1 \\ 0 \end{pmatrix}$$

respectively, in a continuous universe it seems quite natural and mathematically sound to associate semi-true and semi-false expressions with logic vectors of the general form

$$| p > = \begin{pmatrix} \bar{p} \\ p \end{pmatrix}$$

where $\bar{p}$, p are fractional continuous truth-values defined in the unit interval: $0 \leq \bar{p},\ p \leq 1$. In the context of a matrix formulation of logic this makes good theoretical sense as a clarification of approximate reasoning. In the modal probabilistic interpretation the components $\bar{p}$ and p can be viewed as probabilities:

p – as the probability of the true state or state (0,1);

$\bar{p}$ – as the probability of the false state or state (1,0);

and

$$\bar{p} + p = 1 \, .$$

The probability that the logic vector is either in the true or in the false state is 1, the strong normalization condition.

The major advantage of the matrix logic treatment of indeterministic logic is that it allows a unified description, which is able to account equally well for both discrete deterministic logic and continuous indeterministic logic.
The same fundamental bracket device that we employed to define the basis set of logic functions in a discrete universe is also applicable in a continuous universe, with no modification being imposed. Both discrete and continuous logic functions can be computed in a uniform manner, cancelling altogether the need to distinguish between classical and modal logics, which are incorrectly assumed to be two different branches of logic inquiry. In the discrete universe logic functions can yield 4 distinct results, defined in the binary set E_2:

$$\langle x | L | y \rangle \in E_2$$

In the continuous universe logic functions can obtain infinitely many values, defined in the unit interval:

$$0 \leq \langle p | L | q \rangle \leq 1 .$$

In numerical terms the continuous measure for logic functions resembles probability, which is also defined in the unit interval. However, instead of probabilities we shall consider a continuum of fractional truth-values, and use, for a logical function whose arguments are events, the exact same algebraic function of the probabilities of those events. In this way we generalize the notion of event either as a physical event, characterized by probability, or as an event of the mind, characterized by a truth-value.
This enables us to employ positive-valued truth-values as probabilities and probabilities as truth-values. With the extension of truth-values into the non-integer domain we are able to encompass events with approximate truth status. Thus we may not speak of the probabilities of logic expressions, but of the probability of events defined in this universal way.
It has to be emphasized that the concept of truth-value is wider than the concept of probability: while probability can always be interpreted as a truth-value, not all truth-values can be viewed as probabilities. Thus, there are fundamental reasons for interpreting probabilities as fractional truth-values and not the other way round.
The fact that probability is part of logic theory and not vice versa will be further substantiated in the next chapter when we consider the negative truth-values, which are incompatible with the strictly positive measure of probability.

Modal logic functions

Logic functions in the continuous universe can be conveniently presented by the following modal set diagrams:

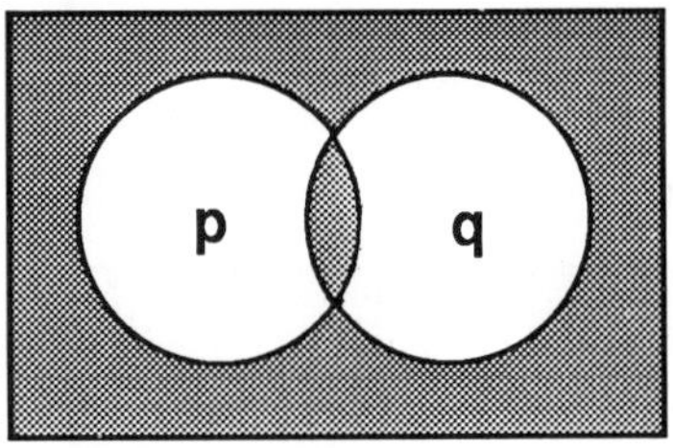

<p | I | q>

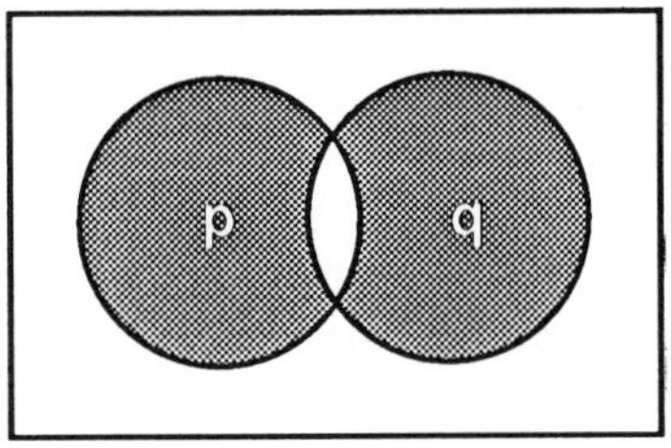

<p | ¬ | q>

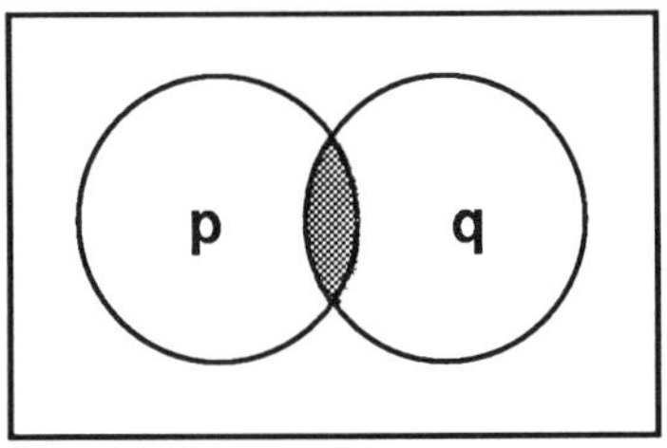

<p | ∧ | q>

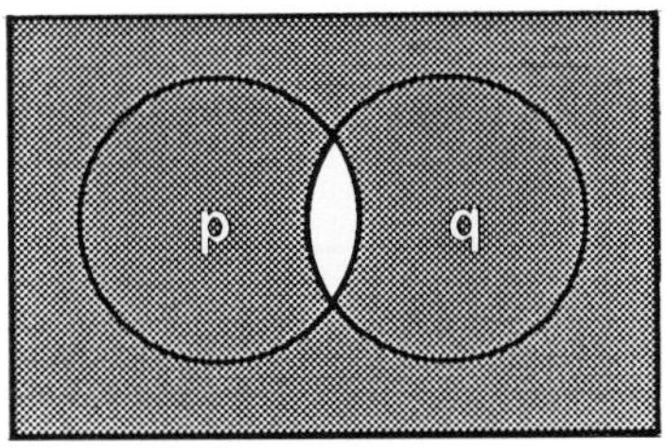

<p | ↓ | q>

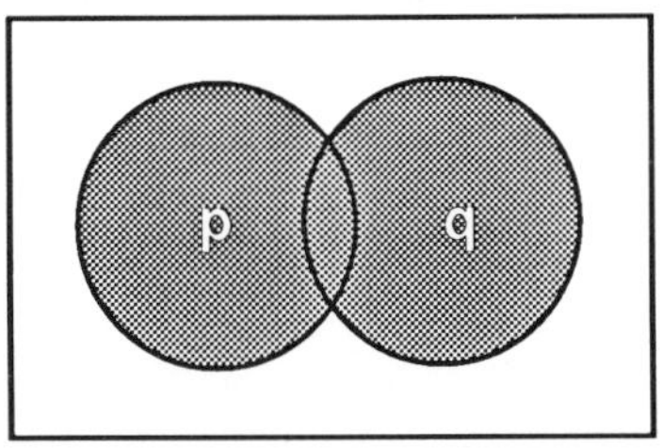

<p | ∨ | q>

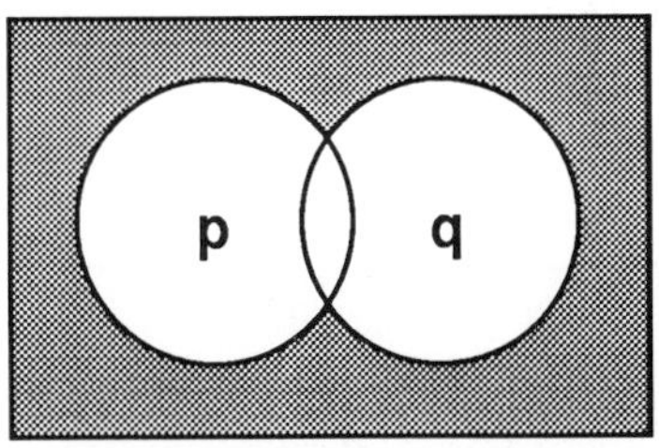

<p | ↑ | q>

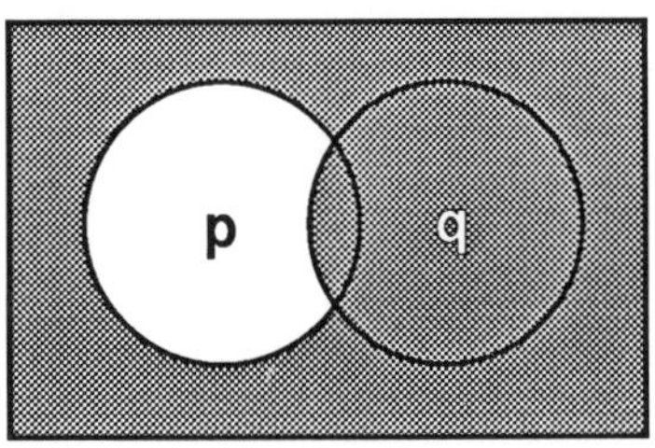

<p | → | q>

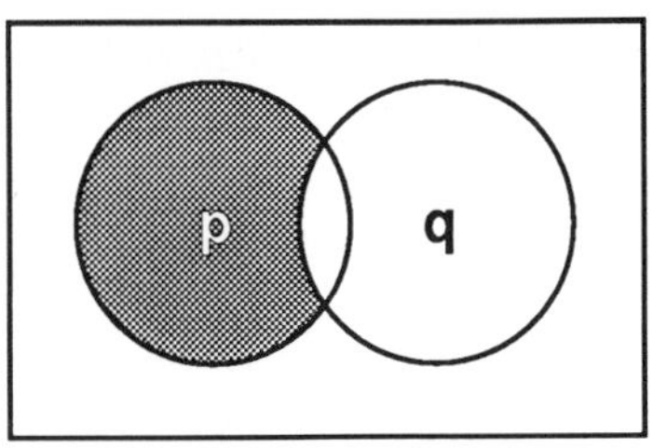

<p |\| q>

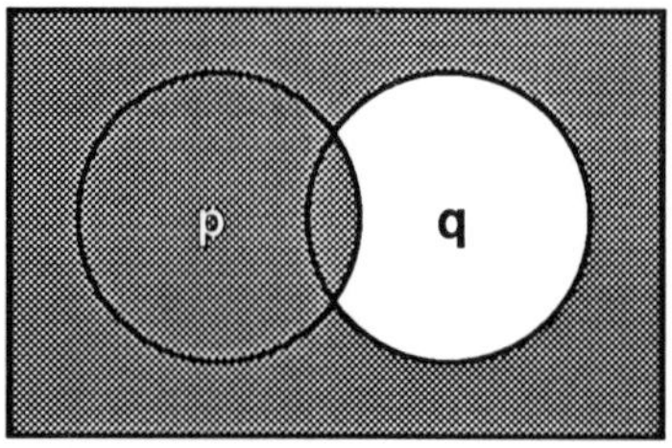

$\langle p|\leftarrow|q\rangle$

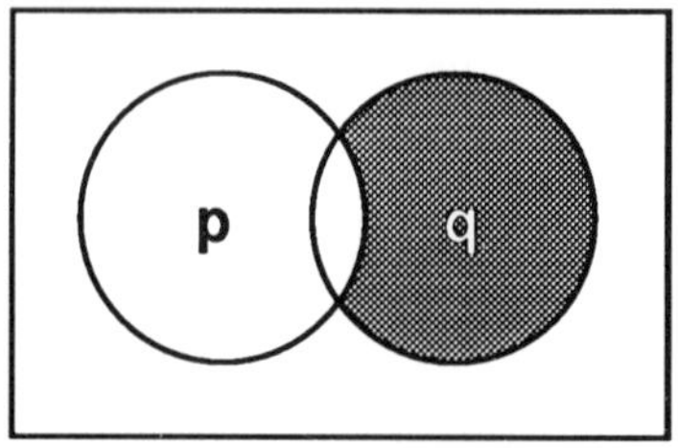

$\langle p|/|q\rangle$

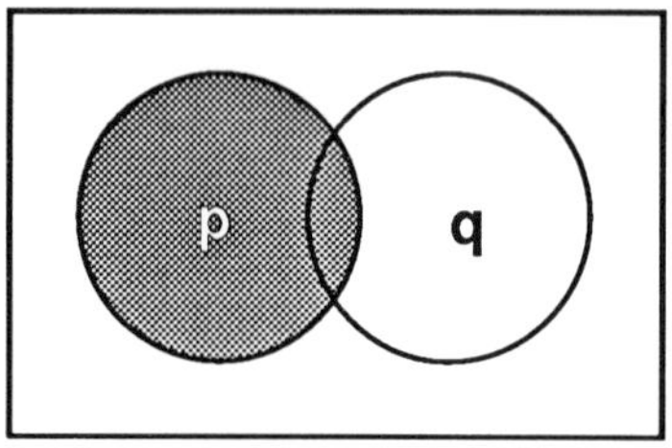

$\langle p|\perp|q\rangle$

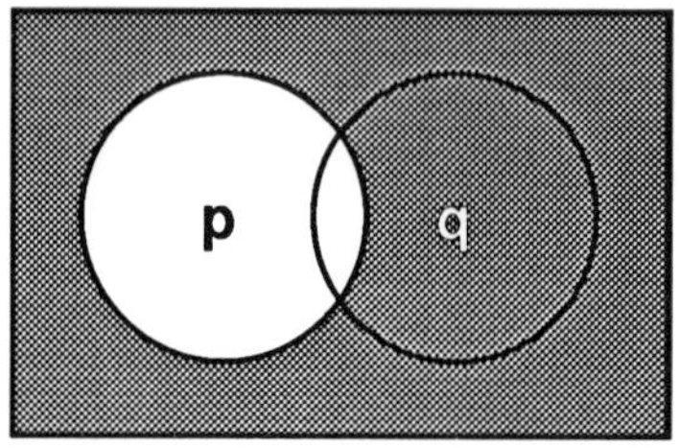

$\langle p|\top|q\rangle$

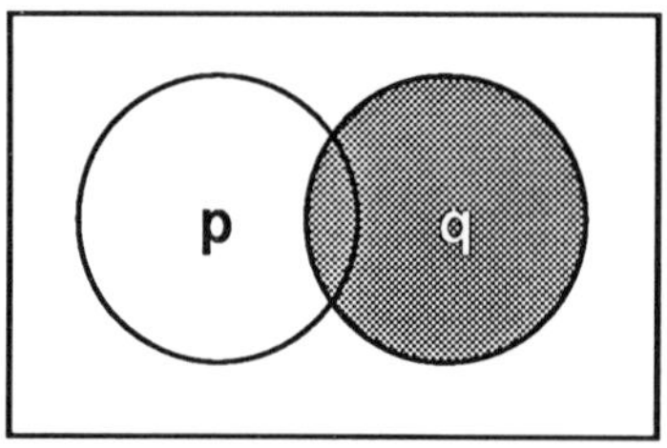

$\langle p|\dashv|q\rangle$

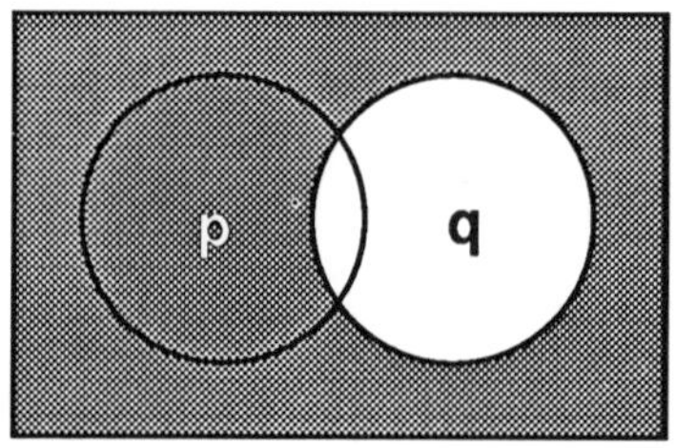

$\langle p|\vdash|q\rangle$

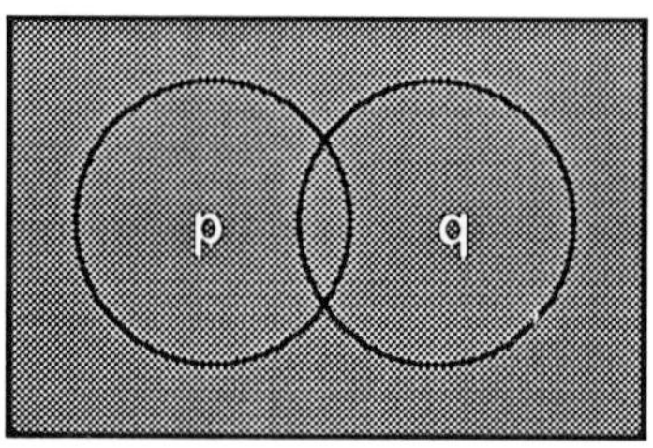

$\langle p|\hat{1}|q\rangle$

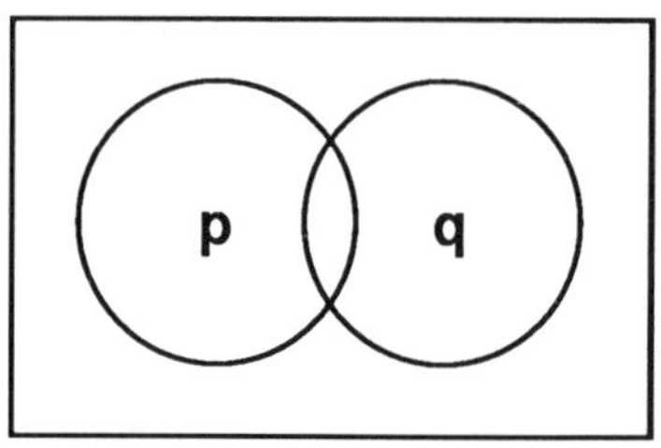

$\langle p|\hat{0}|q\rangle$

The set diagrams above clearly show that the results of evaluation in continuous matrix logic coincide with the results obtained by probability algebra. Compare the expressions in continuous matrix logic with the corresponding expressions in probability logic:

Logic space	Probability logic
$\langle p \mid \wedge \mid q\rangle = pq$	$P(x \wedge y) = P(x)\,P(y)$
$\langle p \mid \vee \mid q\rangle = p + q - pq$	$P(x \vee y) = P(x) + P(y) - P(x \wedge y)$
$\langle p \mid \hat{0} \mid q\rangle = 0$	$P(\varnothing) = P(\bar{\Omega}) = 0$
$\langle p \mid \hat{1} \mid q\rangle = 1$	$P(\Omega) = 1$

If the fractional truth-values are simply renamed as probabilities, the distinction between logic evaluations and probability evaluations eventually vanishes.

However, it is very important to emphasize that while probability logic employs different formulas to evaluate modal conjunction and modal disjunction

$$P(x \wedge y) = P(x)\,P(y)$$
$$P(x \vee y) = P(x) + P(y) - P(x \wedge y)\,,$$

in continuous matrix logic both the conjunction and disjunction functions, in fact any logic function, can be expressed by one universal device $\langle p \mid L \mid q\rangle$, the same atomic device we employed to describe discrete deterministic logic:

$$\langle p \mid \wedge \mid q\rangle = pq$$
$$\langle p \mid \vee \mid q\rangle = p\bar{q} + \bar{p}q + pq = p + q - pq$$

Whereas probability logic employs the laws of probability theory in order to evaluate modal truth-values, matrix logic can obtain the same results on the basis of logic considerations alone. Moreover, a prospect with rich mathematical possibilities emerges where classical probability laws can be derived from the logic system! It follows that matrix logic conjunction and disjunction define the two fundamental laws of probability. This substantiates our claim that the matrix formulation of logic establishes a direct link between logic and probability theory. Because we can pass from probability expressions to logic expressions and vice versa, the treatment of probability and the treatment of logic can be done jointly and not separately. Indeed, the central problems of the theories of logic and probability are quite similar. In probability theory we seek to obtain a probability of a compound event expressed in terms of the probabilities of the constituent events. Similarly, the central problem of logic is to evaluate the truth-value of a compound expression on the basis of the truth-values of atomic expressions. After we identify the probabilities with fractional truth-values, the central problems of both logic and probability can be viewed within the framework of matrix logic as one and the same problem.

Not only do the theories of logic and probability coincide in the essence of their central problem, but matrix logic allows us to join both theories into a single concept computationally. As a result it becomes possible to replace probability estimations with logic transformations, the final outcome of which can be interpreted equally well as either a truth-value or as a probability as the need demands. Conversely, probability methods can be employed for the solution of logic problems. First we pass to a probability expression, perform our transformations and then return to the domain of logic. The fact that the probability estimations can be achieved by logic transformations and vice versa sheds new light on the intuitive nature of the thought processes as opposed to the algorithmic evaluations.

Matrix logic not only joins logic and probability methods, it also allows one to derive both classical and modal logics from the same formal base! The derivation of discrete and modal logics from the same formal base clearly demonstrates the new computing power of matrix logic and provides novel insights into the fundamentals of logic and the intelligence code in general.

Unary operations

The modal properties of logic operations introduce an important new factor into logic analysis, and a large class of theoretical and applied problems can be formulated in terms of the framework of continuous logic.

Just as in the discrete universe, we can consider in the continuous universe unary operations of logic operators, left and right products:

$$L \mid p\rangle \qquad \text{and} \qquad \langle p \mid L \, .$$

Generally the unary operations result in fractional logic vectors which are either normalized or not. The interesting feature of the continuous universe is that, beginning with a fractional-valued logic vector, we can end up with integer-valued undecided vectors:

$$\hat{0} \mid p\rangle = \mid \mathbf{0}\rangle$$
$$\hat{1} \mid p\rangle = \mid \mathbf{1}\rangle$$

or even reach the shell of a logic space, obtaining fully deterministic true or false logic states. This is an advantageous property of the unary application of the filters $\{\bot, \mathrm{T}, \dashv, \vdash\}$:

For left logic:

$$\begin{pmatrix} 0 & 0 \\ 1 & 1 \end{pmatrix}\begin{pmatrix} \bar{p} \\ p \end{pmatrix} = \begin{pmatrix} 0 \\ \bar{p}+p \end{pmatrix} = \begin{pmatrix} 0 \\ 1 \end{pmatrix}$$

$$\begin{pmatrix} 1 & 1 \\ 0 & 0 \end{pmatrix}\begin{pmatrix} \bar{p} \\ p \end{pmatrix} = \begin{pmatrix} \bar{p}+p \\ 0 \end{pmatrix} = \begin{pmatrix} 1 \\ 0 \end{pmatrix}$$

For right logic:

$$\begin{pmatrix} \bar{p} & p \end{pmatrix}\begin{pmatrix} 0 & 1 \\ 0 & 1 \end{pmatrix} = \begin{pmatrix} 0 & \bar{p}+p \end{pmatrix} = \begin{pmatrix} 0 & 1 \end{pmatrix}$$

$$\begin{pmatrix} \bar{p} & p \end{pmatrix}\begin{pmatrix} 1 & 0 \\ 1 & 0 \end{pmatrix} = \begin{pmatrix} \bar{p}+p & 0 \end{pmatrix} = \begin{pmatrix} 1 & 0 \end{pmatrix}$$

<u>Implication</u>

Continuous matrix logic removes the major deficiency of conventional probability logic. This concerns conditional probability, which in probability logic is commonly employed as a model of modal implication:

$$P(q \mid p) = \frac{P(p \wedge q)}{P(p)} = P(q \text{ if } p) \,.$$

In this, probability logic fails to distinguish between conditional probability and the probability of the conditional. The formula gives, correctly, the value of the conditional probability, $P(q \mid p)$, and not the probability of the conditional $\langle p \mid \rightarrow \mid q \rangle$.
In probability logic, if p and q are independent expressions, then:

$$P(q \mid p) = P(q \mid \bar{p})$$

In matrix logic

$$\langle p \mid \rightarrow \mid q \rangle = \langle \bar{p} \mid \rightarrow \mid q \rangle$$

only if

$$\bar{p} = p = \tfrac{1}{2},$$

$$\begin{pmatrix} \bar{p} & p \end{pmatrix}\begin{pmatrix} 1 & 1 \\ 0 & 1 \end{pmatrix} = \begin{pmatrix} p & \bar{p} \end{pmatrix}\begin{pmatrix} 1 & 1 \\ 0 & 1 \end{pmatrix}$$

$$\bar{p} + 1 = p + 1$$

$$\bar{p} = p$$

Whereas the conditional is not defined for P(p) = 0, the implication function is valid for the false antecedent, as one expects it to be. The implication function in the continuous universe is defined without restriction on the value of the antecedent, including its lower bound, p = 0,

$$<p\,|\rightarrow|\,q> \;=\; <0\,|\rightarrow|\,q> \;=$$

$$= (1 \;\; 1)\begin{pmatrix}\bar{q}\\ q\end{pmatrix} = 1\,,$$

just as in the discrete universe Ω□ . *Ex falso quodlibet.*

Tautology

Tautologies have so far been considered in the discrete deterministic domain of truth-values. The very idea of introducing tautology into the continuum of approximate truth-values seems quite improbable, though it is evident that the indeterministic mode of reasoning is an important component of intelligent operations. Contemporary logic offers an adequate account of the deterministic mode of high-level intelligence, that is for tautologies of strictly binary states of mind. Thus, the fundamental problem that arises is to define tautologies that are universally valid.

In view of the fact that high-level intelligence is a covariant phenomenon, the extension of logic into the continuous universe means that the invariance of tautologies has clearly non-trivial consequences.

The question of how invariance can be retained in the indeterministic continuous universe is answered by the unique properties of the tautology operator $\hat{1}$. Of the logic operators, the tautology operator and the empty operator are the only ones that operate identically in either the discrete deterministic universe or the continuous indeterministic universe:

$$<x\,|\,\hat{1}\,|\,y> \;\equiv\; 1\,.$$

$$<p\,|\,\hat{1}\,|\,q> \;=\; \bar{p}\,\bar{q} + \bar{p}\,q + p\,\bar{q} + p\,q \;\equiv\; 1\,.$$

In the continuous universe the tautology function allows us to obtain certainty from any degree of uncertainty. The final certainty arising from the interactions of uncertainties is an important factor which brings new elements into the concept of logic truth and permits us to reconcile the indeterministic roots of intelligence with its deterministic covariant results.

Duality relations

The tautology invariant

$$\bar{p}\,\bar{q} + \bar{p}\,q + p\,\bar{q} + p\,q \equiv 1$$

can be effectively explored for the derivation of duality transformations. The procedure for deriving particular duality laws from tautology is as follows.
Transfer a certain term or terms from one side of the equation to the other. For instance,

$$p\,q = 1 - (\bar{p}\,\bar{q} + \bar{p}\,q + p\,\bar{q})\,.$$

Then by matricization we get

$$\begin{aligned} \langle p | \wedge | q \rangle &= 1 - \langle p | \downarrow | q \rangle = \\ &= \langle \overline{p | \downarrow | q} \rangle\,. \end{aligned}$$

Finally, employing the identity decomposition rule, we derive the duality equation:

$$\langle p | \wedge | q \rangle = \langle \overline{\bar{p} | \vee | \bar{q}} \rangle\,.$$

Consider some other derivations:

(a) $\bar{p}\,q + p\,\bar{q} + p\,q = 1 - \bar{p}\,\bar{q}$

$$\langle p | \vee | q \rangle = \langle \overline{p | \uparrow | q} \rangle = \langle \overline{\bar{p} | \wedge | \bar{q}} \rangle\,.$$

(b) $\bar{p}\,\bar{q} = 1 - (\bar{p}\,q + p\,\bar{q} + p\,q)$

$$\langle p | \uparrow | q \rangle = \langle \overline{p | \vee | q} \rangle = \langle \overline{\bar{p} | \downarrow | \bar{q}} \rangle\,.$$

(c) $(\bar{p}\,\bar{q} + \bar{p}\,q + p\,\bar{q}) = 1 - p\,q$

$$\langle p | \downarrow | q \rangle = \langle \overline{p | \wedge | q} \rangle = \langle \overline{\bar{p} | \uparrow | \bar{q}} \rangle\,.$$

2.3.2 Multiplication table

A transition to the continuous universe not only extends the range of logic values to the entire unit interval, it also, and more importantly, broadens the scope of interactions for logic operators. The multiplication table of continuous logic is presented in Table 2.4:

×	↔	⌉	∧	∨	→	←	↓	↑	\	/	⊥	T	⊣	⊢	**1**	**0**
↔	↔	⌉	∧	∨	→	←	↓	↑	\	/	⊥	T	⊣	⊢	**1**	**0**
⌉	⌉	↔	/	→	∨	↓	←	\	↑	∧	T	⊥	⊣	⊢	**1**	**0**
∧	∧	\	∧	⊥	∧	⊥	\	**0**	\	**0**	⊥	**0**	∧	\	⊥	**0**
∨	∨	←	⊣		∨→		∨↓	\	⊢	∧	**1**	⊥	∨⊣	∨⊢		**0**
→	→	↓	⊣		$\rightarrow^2$		→↓	↑	⊢	/	**1**	T	→⊣	→⊢		**0**
←	←	∨	∧	←∨		$\leftarrow^2$		⊢	\	⊣	⊥	**1**	←⊣	←⊢		**0**
↓	↓	→	/	↓∨		↓←		⊢	↑	⊣	T	**1**	↓⊣	↓⊢		**0**
↑	↑	/	**0**	/	T	↑	T	↑	**0**	/	**0**	T	/	↑	T	**0**
\	\	∧	**0**	∧	⊥	\	⊥	\	**0**	∧	**0**	⊥	∧	\	⊥	**0**
/	/	↑	/	T	/	T	↑	**0**	↑	**0**	T	**0**	/	↑	T	**0**
⊥	⊥	⊥	∧	⊥∨	⊥→	⊥←	⊥↓	\	\	∧	⊥	⊥	⊥⊣	⊥⊢	⊥1	**0**
T	T	T	/	T∨	T→	T←	T↓	↑	↑	/	T	T	T⊣	T⊢	T1	**0**
⊣	⊣	⊢	⊣	**1**	⊣	**1**	⊢	**0**	⊢	**0**	**1**	**0**	⊣	⊢	**1**	**0**
⊢	⊢	⊣	**0**	⊣	**1**	⊢	**1**	⊢	**0**	⊣	**0**	**1**	⊣	⊢	**1**	**0**
1	**1**	**1**	⊣					⊢	⊢	⊣	**1**	**1**	1⊣	1⊢		**0**
0	**0**	**0**	**0**	**0**	**0**	**0**	**0**	**0**	**0**	**0**	**0**	**0**	**0**	**0**	**0**	**0**

Table 2.4 Multiplication table of continuous matrix logic

In comparison with the discrete universe the number of forbidden interactions has now decreased from 49 to 17.

We are particularly concerned with two octets of interactions that are due to the interactions between the four logic operators, ∨, →, ←, ↓, shown explicitly in Table 2.5. In the discrete

	∨	→	←	↓
∨		$\begin{pmatrix} 0 & 1 \\ 1 & 2 \end{pmatrix}$		$\begin{pmatrix} 1 & 0 \\ 2 & 1 \end{pmatrix}$
→		$\begin{pmatrix} 1 & 2 \\ 0 & 1 \end{pmatrix}$		$\begin{pmatrix} 2 & 1 \\ 1 & 0 \end{pmatrix}$
←	$\begin{pmatrix} 0 & 1 \\ 1 & 2 \end{pmatrix}$		$\begin{pmatrix} 1 & 0 \\ 2 & 1 \end{pmatrix}$	
↓	$\begin{pmatrix} 1 & 2 \\ 0 & 1 \end{pmatrix}$		$\begin{pmatrix} 2 & 1 \\ 1 & 0 \end{pmatrix}$	

Table 2.5 Subtable of continuous matrix logic

universe all these interactions were forbidden. In the continuous universe only one of the octets remains forbidden, the other octet of interactions now being allowed with certain restrictions.

The octet of newly admissible interactions comprises four different logic functions. In spite of the non-binary value 2 present in the operators, the corresponding logic functions have a clear logic meaning:

$$<p\begin{pmatrix}0 & 1\\ 1 & 2\end{pmatrix}q> \;=\; p + q$$

$$<p\begin{pmatrix}1 & 2\\ 0 & 1\end{pmatrix}q> \;=\; \overline{p - q}$$

$$<p\begin{pmatrix}1 & 0\\ 2 & 1\end{pmatrix}q> \;=\; \overline{q - p}$$

$$<p\begin{pmatrix}2 & 1\\ 1 & 0\end{pmatrix}q> \;=\; \bar{p} + \bar{q}$$

According to the subtable the new non-binary logic operators can be generated in a multiplicative manner:

$$<p\begin{pmatrix}0 & 1\\ 1 & 2\end{pmatrix}q> \;=\; <p|\vee\!\rightarrow|q> \;=\; <p|\leftarrow\!\vee|q>$$

$$<p\begin{pmatrix}1 & 2\\ 0 & 1\end{pmatrix}q> \;=\; <p|\rightarrow^2|q> \;=\; <p|\downarrow\!\vee|q>$$

$$<p\begin{pmatrix}1 & 0\\ 2 & 1\end{pmatrix}q> \;=\; <p|\leftarrow^2|q> \;=\; <p|\vee\!\downarrow|q>$$

$$<p\begin{pmatrix}2 & 1\\ 1 & 0\end{pmatrix}q> \;=\; <p|\downarrow\!\leftarrow|q> \;=\; <p|\rightarrow\!\downarrow|q>$$

In the additive formulation each operator is a sum of an operator and its dual:

$$<p\begin{pmatrix} 0 & 1 \\ 1 & 2 \end{pmatrix}q> \; = \; <p\,(\vee + \wedge)\,q> \; = \; <p\,|\,\vee + \vee^{*}\,|\,q>$$

$$<p\begin{pmatrix} 1 & 2 \\ 0 & 1 \end{pmatrix}q> \; = \; <p\,(\rightarrow + /)\,q> \; = \; <p\,|\rightarrow + \rightarrow^{*}|\,q>$$

$$<p\begin{pmatrix} 1 & 0 \\ 2 & 1 \end{pmatrix}q> \; = \; <p\,(\leftarrow + \backslash)\,q> \; = \; <p\,|\leftarrow + \leftarrow^{*}|\,q>$$

$$<p\begin{pmatrix} 2 & 1 \\ 1 & 0 \end{pmatrix}q> \; = \; <p\,(\downarrow + \uparrow)\,q> \; = \; <p\,|\downarrow + \downarrow^{*}|\,q>$$

In the continuous universe the operation of the addition of logic connectives gains a new and constructive logic meaning in that a significant number of new compound logic connectives can be constructed.

An operator sum like

$$\vee + \wedge$$

is not a formal addition of matrices but a compound connective operator with a definite logic interpretation in continuous matrix logic.

The interesting property of the compound connective operators,

$$(\vee + \wedge), \quad (\rightarrow + /), \quad (\vee + \backslash), \quad (\downarrow + \uparrow)\,,$$

is that in a continuous universe with equilibrium, when all truth-values equal $\frac{1}{2}$, the corresponding logic functions are all true in a deterministic sense:

$$\begin{aligned} <p\,|\,\vee + \wedge\,|\,q> &= \; <p\,|\rightarrow + /\;|\,q> \; = \\ &= \; <p\,|\,\vee + \backslash\;|\,q> \; = \\ &= \; <p\,|\downarrow + \uparrow|\,q> = 1\,, \\ p = q = \tfrac{1}{2}\,. & \end{aligned}$$

The important result of the study of logic interactions in the continuous universe is the fundamental extension of the integer alphabet of truth-values. The extended alphabet of logic is now as follows:

$\{0, 1, 2\}$

0 — false
1 — true
2 — double true

The analysis of the multiplicative and additive interactions in the continuous universe has shown that the number of admissible interactions has now, in comparison with discrete logic, grown significantly. This indicates a fundamental trade-off occurring in continuous logic. The weakening of strictly binary values into a continuum of logic values cancels a number of previously forbidden interactions and allows one to construct new logic connective operators, significantly increasing the expressive power of logic. This trade-off provides an important clarification of the creative and artistic capabilities of the abstract mind, which tolerates ambiguity and imprecision. The creative power of the intellect appears to be inversely related to the extent of the deterministic component. The less Boolean the logic, the more creative the mind.

2.3.3 Correction terms

A major difference between discrete and continuous logics is reflected in their compliance with the idempotency law. Whereas the discrete universe is involutive:

$$x^2 = x ,$$

in the continuous universe:

$$p^2 \leq p ,$$

where the equality holds only in the limiting cases.
Consequently, in discrete logic the absorption rule applies to identical expressions which can be cancelled with no effect on the resulting value of conjunction. But in continuous logic such cancellation is in general not possible. The identical expressions may have a cumulative effect whereby the resulting uncertainty increases with the increase in the number of identical expressions. This clearly alters fundamentally the structure of the contradiction law in the continuous universe: we have a weak contradiction law,

$$p\bar{p} \leq 0 ,$$

because the product of the complementary expressions is not necessarily empty. The equality holds only in the limiting cases. This allows us to conclude that deterministic logic is the binary limit of continuous logic. For a given logic function a transition to the discrete universe is obtained by a limiting process:

$$< p \mid L \mid q > \underset{p,\, q \rightarrow E_2}{\Rightarrow} < x \mid L \mid y >$$

Here the fractional truth-values converge to the binary set $E_2 = \{0, 1\}$. Conversely, a discrete logic function is changed into a continuous logic function if binary truth-values are relaxed into fractional values. A transition from a deterministic logic function to a corresponding continuous logic function is obtained simply by replacing integer truth-values by fractional values:

$$<x \mid L \mid y> \;\Rightarrow\; <p \mid L \mid q> .$$

Everything is quite simple for logic functions of two variables. For compound logic functions this transition becomes more complex, requiring the introduction of correction terms.

Transmultiplication

We introduce correction terms for the transmultiplication fusion of logic operators. In the discrete logic $\Omega^{\square}$ the equivalence function can be expressed as a compound function:

$$<x \mid y> \;=\; <x \mid \rightarrow \mid y> <x \mid \leftarrow \mid y> .$$

With a transition to the continuous logic $\Omega^{\Diamond}$ the left side of the equality must be changed into $<p \mid q>$, but

$$<p \mid q> \;\neq\; <p \mid \rightarrow \mid q> <p \mid \leftarrow \mid q> .$$

The transmultiplication law does not hold in the continuous universe and fusion of implications does not occur. In order to re-establish the equality we require a correction term:

$$<p \mid q> \;=\; <p \mid \rightarrow \mid q> <p \mid \leftarrow \mid q> - <p \mid \backslash \mid q> <p \mid / \mid q> .$$

The general formula for the correction term of transmultiplication is as follows:

if $\quad L = L' \circledast L''$, then

$$< p \mid L \mid q > \;=\; < p \mid L' \mid q > < p \mid L'' \mid q > - < \overline{p \mid L' \mid q} > < \overline{p \mid L'' \mid q} > .$$

Consider these correction terms:

$$\text{(a)}\quad <\overline{p \,|\, q}> \;=\; <p\,|\vee|\,q><p\,|\!\downarrow\!|\,q> - <\overline{p\,|\vee|\,q}><\overline{p\,|\!\downarrow\!|\,q}> \;=$$

$$= \bar{\bar{p}}\,\bar{\bar{q}}\,\overline{p\,q} - \bar{p}\,\bar{q}\,p\,q = 1 - \bar{p}\,\bar{q} - p\,q + \bar{p}\,\bar{q}\,p\,q - \bar{p}\,\bar{q}\,p\,q =$$

$$= 1 - \bar{p}\,\bar{q} - p\,q = p\,\bar{q} + \bar{p}\,q = <\overline{p \,|\, q}>$$

$$\text{(b)}\quad <1 \,|\, q> \;=\; <p \vee q><p \rightarrow q> - <\overline{p \vee q}><\overline{p \rightarrow q}> \;=$$

$$= \bar{\bar{p}}\,\bar{\bar{q}}\,\overline{p\,\bar{q}} - \bar{p}\,\bar{q}\,p\,\bar{q} = 1 - \bar{p}\,\bar{q} - p\,\bar{q} + \bar{p}\,\bar{q}\,p\,\bar{q} - \bar{p}\,\bar{q}\,p\,\bar{q} =$$

$$= <\overline{p\,|\vdash|\,q}> \;=\; <p\,|\dashv|\,q> \;=\; <1\,|\,q>$$

$$\text{(c)}\quad 0 = <p\,|\hat{0}|\,q> = <p\,|\wedge|\,q><p\,|\!\downarrow\!|\,q> - <\overline{p\,|\wedge|\,q}><\overline{p\,|\!\downarrow\!|\,q}> =$$

$$= p\,q\,\overline{p\,q} - \overline{p\,q}\,p\,q = 0$$

With a transition from continuous logic to discrete logic the transmultiplication law is re-established in its original form and the correction term converges to zero.

2.3.4 Universum operator

Matrix logic makes it possible to solve a wide range of problems that are inaccessible to other forms of logic. We have determined earlier that the division into operators and operands employed in conventional logic is in fact an artificial device, stemming from the incomplete formalization of conventional logic.

In matrix logic, a logic operator and logic vectors are always interchangeable. In order to carry out the interchange we express a logic operator as the outer product of logic vectors and then replace the vectors of the operator with those of the operands:

$$<x_1\,|\,L'\,|\,y_1> = <x_1\,|\,x_2><y_2\,|\,y_1> = <x_2\,|\,x_1><y_1\,|\,y_2> = <x_2\,|\,L''\,|\,y_2>\,.$$

In the discrete logic universe this simple, yet fundamental, mutation is fully symmetric: both sets of logic vectors in operator and operands are from the same class of binary-valued logic vectors.

A mutation of a deterministic device $<x_1 \,|\, L' \,|\, y_1>$ into the equivalent device $<x_2 \,|\, L'' \,|\, y_2>$ introduces no new features into the logic process and may pass unnoticed. Though the operator L'' may differ from the operator L', they always both belong to the same basic set of logic operators.

The strong symmetry that characterized the discrete logic space is no longer present in the continuous universe of matrix logic.
Consider a general form of a logic function

$$<p \mid L \mid q>$$

where L is a logic operator, p, q are non-binary logic vectors. Employing the outer product representation for the operator L, we can obtain a transformation

$$<p \mid L \mid q> \;=\; <x \mid p><q \mid y> \;=\; <x \mid \Omega \mid y>$$

where $\Omega = \mid p><q \mid$, $<x \mid$, $\mid y>$ are binary-valued logic vectors, including both the basis vectors and the undecided vectors as well,

$$<\mathbf{1} \mid \;=\; (1 \quad 1)$$
$$<\mathbf{0} \mid \;=\; (0 \quad 0)\,.$$

In reverse, the outer product operator

$$\Omega \;=\; \mid p><q \mid \;=\; \begin{pmatrix} \bar{p}\,\bar{q} & \bar{p}\,q \\ p\,\bar{q} & p\,q \end{pmatrix}$$

is in the general case represented by a matrix with non-integer elements.
The analysis of the properties of the outer product matrix Ω indicates that Ω represents an invariant quantity and can be defined as the universum logic operator. The universality of the operator Ω resides in the fact that it allows us to express any logic function in a uniform fashion.
Applying binary logic vectors to the operator Ω, we can obtain the logic functions of the operators of the logic base

$$<1 \mid \Omega \mid 1> \;=\; <p \mid \wedge \mid q>$$

$$<1 \mid \Omega \mid 0> \;=\; <p \mid \backslash \mid q>$$

$$<0 \mid \Omega \mid 1> \;=\; <p \mid / \mid q>$$

$$<0 \mid \Omega \mid 0> \;=\; <p \mid \uparrow \mid q>$$

A rotation of a logic space may change the logic vectors $<p \mid$ and $\mid q>$ which participate in the outer product generation of the universum logic operator Ω. Consequently, the individual components of Ω may change in value. Independently of particular values obtained by $\langle p|$ and $|q\rangle$, they would, however, jointly still characterise the same logic quantity: the norm, that is, the total sum of the components of the operator Ω, is invariably a tautology:

$$\|\Omega\| = \bar{p}\,\bar{q} + \bar{p}\,q + p\,\bar{q} + p\,q \equiv 1 .$$

The norm convolution of the operator Ω is a fundamental logic invariant, which is most clearly seen in set diagram formulation:

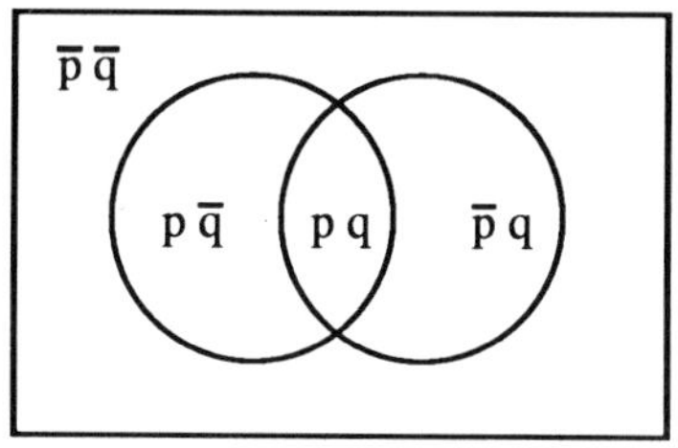

The components of Ω divide the complete tautology field into four regions, the sum of which is invariably a tautology.

The representation of a logic function via the universum operator Ω may require both a base logic vector and an undecided vector. This is the case with the filter, antifilter and their converses:

$$\langle p | \bot | q\rangle = \langle 1 | \Omega | \mathbf{1}\rangle$$

$$\langle p | \mathrm{T} | q\rangle = \langle 0 | \Omega | \mathbf{1}\rangle$$

$$\langle p | \dashv | q\rangle = \langle \mathbf{1} | \Omega | 1\rangle$$

$$\langle p | \vdash | q\rangle = \langle \mathbf{1} | \Omega | 0\rangle$$

For the contradiction and tautology functions the logic operators are decomposed into a pair of undecided vectors:

$$\langle p | \hat{0} | q\rangle = \langle \mathbf{0} | \Omega | \mathbf{0}\rangle$$

$$\langle p | \hat{1} | q\rangle = \langle \mathbf{1} | \Omega | \mathbf{1}\rangle$$

For four logic functions realization by the universum logic operator Ω involves the complement operator:

$$\langle p | \vee | q \rangle = \langle \overline{0 | \Omega | 0} \rangle$$

$$\langle p | \rightarrow | q \rangle = \langle \overline{1 | \Omega | 0} \rangle$$

$$\langle p | \leftarrow | q \rangle = \langle \overline{0 | \Omega | 1} \rangle$$

$$\langle p | \downarrow | q \rangle = \langle \overline{1 | \Omega | 1} \rangle$$

And finally, a single bracket device is insufficient for the realization of two fundamental functions: equivalence and negation. The resulting expression here takes the form of an algebraic sum of two bracket products:

$$\langle p | \leftrightarrow | q \rangle = \langle 1 | \Omega | 1 \rangle + \langle 0 | \Omega | 0 \rangle$$

$$\langle p | \neg | q \rangle = \langle 1 | \Omega | 0 \rangle + \langle 0 | \Omega | 1 \rangle$$

The exchange of binary-valued operators and non-integer logic vectors, leading to the formation of the universum logic operator Ω, represents an important transformation. Because the deterministic operators are replaced by the probabilistic operator Ω, while at the same time the probabilistic logic vectors are replaced by binary logic vectors, the overall metamorphosis can be viewed as a transition from probability logic space to deterministic logic space.
If required, matrix logic allows one to pass from probability to certainty and vice versa. This involves replacing discrete integer operators with the non-integer universum operator Ω. This has direct parallels with two important problems in psychology that arise in the study of the intellect:

$$\langle x | ? | y \rangle \quad \text{and} \quad \langle ? | L | ? \rangle .$$

The theory of the universum operator Ω makes it possible to pass from one to the other easily.

Trace

The important characteristic of the universum logic operator Ω is its trace. The trace Tr Ω is expressed as the logic function of equivalence:

$$\mathrm{Tr}\,\Omega = \bar{p}\,\bar{q} + p\,q = \langle p | q \rangle$$

The formula is a consequence of the tensor operation of transvection:

$$\langle p | q \rangle = \mathrm{Tr}\, | p \rangle \langle q |$$

Taking into account the fundamental tautology relation, Tr Ω can be expressed in terms of the secondary diagonal as follows

$$\mathrm{Tr}\,\Omega = \overline{\bar{p}\,q + p\,\bar{q}}\,.$$

This gives us the valuable rule of the elimination of the complement operator:

$$\overline{\mathrm{Tr}\,|\,p><q\,|} = \mathrm{Tr}\,|\,\bar{p}><q\,| = \mathrm{Tr}\,|\,p><\bar{q}\,|\,.$$

In logic theory the trace Ω is restricted to the unit interval,

$$0 \leq \mathrm{Tr}\,\Omega \leq 1\,,$$

where the boundary values are obtained only in the limiting deterministic cases:

$\mathrm{Tr}\,\Omega = 1$ — for collinear logic vectors
$\mathrm{Tr}\,\Omega = 0$ — for orthogonal logic vectors

In the case of collinearity, true and false are not distinguishable by their traces, which are the same.

Complementation

The logic built around the universum logic operator Ω is singular. In logic terms, the determinant $|\,\Omega\,|$ is necessarily a contradiction:

$$\det\,\Omega = \begin{vmatrix} \bar{p}\,\bar{q} & \bar{p}\,q \\ p\,\bar{q} & p\,q \end{vmatrix} = \bar{p}\,\bar{q}\,p\,q - p\,\bar{q}\,\bar{p}\,q = 0\,.$$

Complementation can transform Ω into a non-singular operator $\overline{\Omega}$:

$$\det\,\overline{\Omega} = \begin{vmatrix} \overline{\bar{p}\,\bar{q}} & \overline{\bar{p}\,q} \\ \overline{p\,\bar{q}} & \overline{p\,q} \end{vmatrix} = \overline{\bar{p}\,\bar{q}}\;\overline{p\,q} - \overline{p\,\bar{q}}\;\overline{\bar{p}\,q} =$$

$$= 1-\bar{p}\,\bar{q}-p\,q+\bar{p}\,\bar{q}\,p\,q- 1+\bar{p}\,q+p\,\bar{q}-\bar{p}\,q\,p\,\bar{q} =$$

$$= \bar{p}\,q+p\,\bar{q}-\bar{p}\,\bar{q}-p\,q =$$

$$= <\overline{p\,|\,q}>-<p\,|\,q>\,.$$

The determinant $|\overline{\Omega}|$ represents the difference of two logic functions:

$$\det \overline{\Omega} \;=\; \langle \overline{p \mid q} \rangle - \langle p \mid q \rangle \;.$$

In terms of the traces of the operator Ω this can be expressed as follows:

$$\det \overline{\Omega} \;=\; \overline{\mathrm{Tr}\;\Omega} \;-\mathrm{Tr}\;\Omega\,.$$

Whenever the trace of the operator Ω equals its complement,

$$\overline{\mathrm{Tr}\;\Omega} \;=\; \mathrm{Tr}\;\Omega\,,$$

the complement of the universum operator $\overline{\Omega}$ is singular.
Because

$$\overline{\mathrm{Tr}\;\Omega} \;+\; \mathrm{Tr}\;\Omega \;=\; 1\,,$$

the singularity condition is

$$\mathrm{Tr}\,\Omega \;=\; \frac{1}{2}\,,$$

which in turn implies

$$p \;=\; q \;=\; \frac{1}{2}\,.$$

Since

$$\mu\;(\Omega) \;=\; 1\,,$$

the complement of each component of an operator Ω is equal to the sum of the remaining components (cofactor):

$$\overline{\begin{pmatrix} \bar{p}\,\bar{q} & \bar{p}\,q \\ p\,\bar{q} & p\,q \end{pmatrix}} = \begin{pmatrix} \overline{\bar{p}\,\bar{q}} & \overline{\bar{p}\,q} \\ \overline{p\,\bar{q}} & \overline{p\,q} \end{pmatrix} = \begin{pmatrix} 1-\bar{p}\,\bar{q} & 1-\bar{p}\,q \\ 1-p\,\bar{q} & 1-p\,q \end{pmatrix} =$$

$$= \begin{pmatrix} \bar{p}\,q+p\,\bar{q}+p\,q & \bar{p}\,\bar{q}+p\,\bar{q}+p\,q \\ \bar{p}\,\bar{q}+\bar{p}\,q+p\,q & \bar{p}\,\bar{q}+\bar{p}\,q+p\,\bar{q} \end{pmatrix}.$$

Complementation of the logic operator $\overline{\Omega}$ was employed for the expression of four logic functions:

$$<p \mid \vee \mid q>, \quad <p \mid \rightarrow \mid q>, \quad <p \mid \leftarrow \mid q>, \quad <p \mid \downarrow \mid q> .$$

Employing the operator $\overline{\Omega}$, we can merge these logic functions into a single expression:

$$< x \mid \overline{\Omega} \mid y > = < x \begin{pmatrix} <p \mid \vee \mid q> & <p \mid \leftarrow \mid q> \\ <p \mid \rightarrow \mid q> & <p \mid \downarrow \mid q> \end{pmatrix} y >$$

Similarly for $\{\wedge \setminus / \uparrow\}$:

$$< x \mid \Omega \mid y > = < x \begin{pmatrix} <p \uparrow q> & <p \, / \, q> \\ <p \setminus q> & <p \wedge q> \end{pmatrix} y >$$

<u>Matrix product</u>

Consider the interaction of two universum operators

$$\Omega_{pq} = \mid p> <q \mid$$

and

$$\Omega_{rs} = \mid r> <s \mid .$$

The interaction $\Omega_{pq}\Omega_{rs}$,

$$\begin{aligned} \Omega_{pq}\Omega_{rs} &= \mid p > < q \mid r > < s \mid = \\ &= < q \mid r > \mid p > < s \mid = \\ &= < q \mid r > \begin{pmatrix} \overline{p}\,\overline{s} & \overline{p}\, s \\ p\, \overline{s} & p\, s \end{pmatrix} = \\ &= < q \mid r > \Omega_{ps} \ , \end{aligned}$$

connects the logic vectors $\mid p>$ and $<s \mid$ into a new universum operator Ω_{ps} and expels forward the inner product formed from the remaining two logic vectors $<q \mid$ and $\mid r>$.

Conversely, if a scalar multiplier can be expressed as an inner product of two logic vectors $<q \mid r>$, its product with a universum operator can be translated into the interaction of two universum operators:

$$<q \mid r> \Omega_{ps} = <q \mid r> \mid p> <s \mid = \mid p> <q \mid r> <s \mid = \Omega_{pq} \Omega_{rs}$$

Eigenequation

Consider the eigenequation for a universum operator:

$$\Omega \mid i> = \lambda \mid i>,$$

where $\Omega = \mid p> <q \mid$, $\mid i>$ is an unknown eigenvector, and λ an unknown eigenvalue. Employing a standard technique, we write a characteristic equation:

$$\mid \Omega - \lambda I \mid = 0$$

$$\begin{vmatrix} \bar{p}\,\bar{q} - \lambda & \bar{p}\,q \\ p\,\bar{q} & p\,q - \lambda \end{vmatrix} = (\bar{p}\,\bar{q} - \lambda)(p\,q - \lambda) - p\,\bar{q}\,\bar{p}\,q = 0$$

$$\lambda^2 - \lambda(\bar{p}\,\bar{q} + p\,q) = 0,$$

$$\lambda_1 = 0; \qquad \lambda_2 = \bar{p}\,\bar{q} + p\,q = <p \mid q>$$

Substituting the essential λ_2 into the eigenequation, we calculate the eigenvector $\mid i>$,

$$\mid i> = \mid p>.$$

This completes our objective, and thus the eigenequation for the universum operator is as follows:

$$\Omega \mid p> = \mathrm{Tr}\, \Omega \mid p>,$$

where $\mathrm{Tr}\, \Omega = <p \mid q>$.

We could have avoided this relatively long derivation path. For a universum operator Ω the fundamental eigenvalue is its own trace Tr Ω and its own generating vectors are the eigenvectors which follow directly from the equations:

Right eigenequation:

$$| p\rangle \langle q | p\rangle = \langle p | q\rangle | p\rangle$$

$$\Omega | p\rangle = \mathrm{Tr}\, \Omega | p\rangle$$

Left eigenequation:

$$\langle q | p\rangle \langle q | = \langle q | \langle p | q\rangle$$

$$\langle q | \Omega = \langle q | \mathrm{Tr}\, \Omega$$

Once again we observe that the trace Tr Ω is a fundamental logic quantity. The universum operator possesses remarkable properties. Interaction of Ω with itself induces a homogeneous contraction of logic space:

$$\Omega^2 = | p\rangle \langle q | p\rangle \langle q | = \langle p | q\rangle | p\rangle \langle q |$$

This can be viewed as the eigenequation for the universum operator itself:

$$\Omega\, \Omega = \mathrm{Tr}\, \Omega\, \Omega ,$$

where Tr Ω is an eigenvalue and Ω manifests itself in dual roles as an operator and as an eigenstate matrix, a generalized vector in matrix vector space.

While the logic norm of Ω is tautology,

$$\| \Omega \| \equiv 1 ,$$

the norm of the square of the universum operator equals its trace:

$$\| \Omega^2 \| \equiv \mathrm{Tr}\, \Omega .$$

By induction we obtain

$$\| \Omega^n \| \equiv (\mathrm{Tr}\, \Omega)^{n-1}.$$

Clearly,

$$\mathrm{Tr}\, \Omega^2 = (\mathrm{Tr}\, \Omega)^2,$$

and by induction we obtain

$$\mathrm{Tr}\, \Omega^n = (\mathrm{Tr}\, \Omega)^n.$$

The trace of the universum operator is at the same time its fundamental eigenvalue, and is also a logic norm of the universum autointeraction:

$$\mathrm{Tr}\, \Omega = \langle p \mid q \rangle = \| \Omega^2 \| .$$

Autointeraction causes logic space to contract and because this contraction is spread evenly through the logic space, Ω is an eigenstate with respect to itself.
This result has fundamental implications in the framework of the logic theory of intelligence: a self-interacting evolving intellect will scale itself down. We obtain a continuously contracting universe:

$$\Omega^n = (\mathrm{Tr}\, \Omega)^{n-1}\, \Omega .$$

For consecutive autointeractions, with n converging to infinity, Ω^n converges to the empty operator:

$$\Omega^n \xrightarrow[n \to \infty]{} \hat{0} .$$

Mathematical résumé

The matrix operator formulation of logic which we have presented in this chapter reveals new fundamental properties of logic space and provides important clues for the deciphering of the mechanism of the mind. In order to compare the main features of discrete and continuous logics we reiterate in concise format the principal mathematical relations:

Discrete universe	Continuous universe
Inner products	
$\langle x\mid 1\rangle = x;\ \langle x\mid 0\rangle = \bar{x}$	$\langle p\mid 1\rangle = p;\ \langle p\mid 0\rangle = \bar{p}$
$\langle x\mid y\rangle = \langle y\mid x\rangle$	$\langle p\mid q\rangle = \langle q\mid p\rangle$
$\langle x\mid x\rangle = \langle \bar{x}\mid \bar{x}\rangle = \langle \overline{\overline{x\mid x}}\rangle = 1$	$\langle p\mid p\rangle = \langle \bar{p}\mid \bar{p}\rangle = \langle \overline{\overline{p\mid p}}\rangle \leq 1$
$\langle \overline{x\mid x}\rangle = \langle \bar{x}\mid x\rangle = \langle x\mid \bar{x}\rangle = 0$	$\langle \overline{p\mid p}\rangle = \langle \bar{p}\mid p\rangle = \langle p\mid \bar{p}\rangle \neq 0$
$\langle x\mid y\rangle^2 = \langle x\mid y\rangle$	$\langle p\mid q\rangle^2 \leq \langle p\mid q\rangle$
Outer products	
$\mid x\rangle\langle y\mid\ = L$	$\mid p\rangle\langle q\mid\ = \Omega$
$\mid y\rangle\langle x\mid\ = \tilde{L}$	$\mid q\rangle\langle p\mid\ = \tilde{\Omega}$
$\mid \overline{x\rangle\langle y}\mid\ = \hat{1} - L$	$\mid \overline{p\rangle\langle q}\mid\ = \hat{1} - \Omega$
$(\mid x\rangle\langle x\mid)^2 = \mid x\rangle\langle x\mid$	$(\mid p\rangle\langle p\mid)^2 \neq \mid p\rangle\langle p\mid$
$(\mid x\rangle\langle \bar{x}\mid)^2 = (\mid \bar{x}\rangle\langle x\mid)^2 = \hat{0}$	$(\mid p\rangle\langle \bar{p}\mid)^2 \neq (\mid \bar{p}\rangle\langle p\mid)^2 \neq \hat{0}$
$\Vert \mid x\rangle\langle y\mid \Vert = 1$	$\Vert \Omega \Vert = 1$
$\mathrm{Tr}\mid x\rangle\langle y\mid\ = \delta_{xy}$	$\mathrm{Tr}\,\Omega = \langle p\mid q\rangle$
Bracket products	
$\langle \overline{x\mid L\mid y}\rangle = \langle x\mid \bar{L}\mid y\rangle$	$\langle \overline{p\mid L\mid q}\rangle = \langle p\mid \bar{L}\mid q\rangle$
$\langle x\mid \tilde{L}\mid y\rangle = \langle y\mid L\mid x\rangle$	$\langle p\mid \tilde{L}\mid q\rangle = \langle q\mid L\mid p\rangle$
$\langle x\mid \overline{\tilde{L}}\mid y\rangle = \langle \bar{y}\mid L\mid \bar{x}\rangle$	$\langle p\mid \overline{\tilde{L}}\mid q\rangle = \langle \bar{q}\mid L\mid \bar{p}\rangle$
$\langle x\mid L\mid y\rangle = \langle \bar{x}\mid \theta L\mid \bar{y}\rangle$	$\langle p\mid L\mid q\rangle = \langle \bar{p}\mid \theta L\mid \bar{q}\rangle$
$\langle x\mid L^*\mid y\rangle = \langle \bar{x}\mid \bar{L}\mid \bar{y}\rangle$	$\langle p\mid L^*\mid q\rangle = \langle \bar{p}\mid \bar{L}\mid \bar{q}\rangle$
$\langle x\mid L\mid y\rangle = \langle \bar{x}\mid \widetilde{\overline{\theta L}}\mid \bar{y}\rangle$	$\langle p\mid L\mid q\rangle = \langle \bar{p}\mid \widetilde{\overline{\theta L}}\mid \bar{q}\rangle$
Fundamental products	
$\sum^{m}\wedge^{n} \langle x_i\mid L_j\mid y_k\rangle \in E_2$	$0 \leq \sum^{m}\wedge^{n} \langle p_i\mid L_j\mid q_k\rangle \leq 1$
$\langle x\mid L\mid y\rangle\langle x\mid \bar{L}\mid y\rangle \equiv 0$	$\langle p\mid L\mid q\rangle\langle p\mid \bar{L}\mid q\rangle \neq 0$
$\langle x\mid L\mid y\rangle + \langle x\mid \bar{L}\mid y\rangle \equiv 1$	$\langle p\mid L\mid q\rangle + \langle p\mid \bar{L}\mid q\rangle \equiv 1$

2.3.5 Differential logic

The closure of the gap between logic and mathematics is most clearly apparent in the differential analysis of logic functions. The latter reveals new fundamental properties of logic and opens a principally new way of examining logic operations and processes.

Logic derivatives

It is usual to discuss logic within the framework of discrete mathematics. There are no apparent links with analysis and differential calculus. Within the scope of traditional logic it is inconceivable to think of the differentiation or integration of logic quantities.
Continuous matrix logic, as developed in this chapter, paves the way for the analytic treatment of logic problems to which differential methods become applicable. In Table 2.6 we present the derivatives for the basic set of 16 logic functions.

The derivatives of the logic functions are linear functions or constants. The derivative L(p,q)' of a logic function and the derivative of its complement $\overline{L(p,q)}'$ differ in sign:

$$\overline{L(p,q)}' = -L(p,q)' .$$

The equality of a derivative to 0 indicates that a logic function is either a tautology or a contradiction,

$$<p\,|\,\hat{1}\,|\,q>' = 0, \qquad <p\,|\,\hat{0}\,|\,q>' = 0 ,$$

or is not dependent on a variable of differentiation:

$$<p\,|\bot|\,q>'_q = 0 ; \quad <p\,|\,\mathrm{T}\,|\,q>'_q = 0 ;$$

$$<p\,|\dashv|\,q>'_p = 0 ; \quad <p\,|\vdash|\,q>'_p = 0 .$$

Whereas the logic functions L(p,q) are defined in the unit interval

$$0 \leq L(p,q) < 1 ,$$

the values of their derivatives fall into the balanced interval

$$-1 \leq L\,(p,q)' < 1 .$$

α	$\frac{d\alpha}{dp}$	$\frac{d\alpha}{dq}$
$\langle p \mid \leftrightarrow \mid q \rangle$	$2q - 1$	$2p - 1$
$\langle p \mid \rceil \mid q \rangle$	$1 - 2q$	$1 - 2p$
$\langle p \mid \wedge \mid q \rangle$	q	p
$\langle p \mid \vee \mid q \rangle$	$\bar{q}$	$\bar{p}$
$\langle p \mid \rightarrow \mid q \rangle$	$-\bar{q}$	p
$\langle p \mid \leftarrow \mid q \rangle$	q	$-\bar{p}$
$\langle p \mid \downarrow \mid q \rangle$	$-q$	$-p$
$\langle p \mid \uparrow \mid q \rangle$	$-\bar{q}$	$-\bar{p}$
$\langle p \mid \backslash \mid q \rangle$	$\bar{q}$	$-p$
$\langle p \mid / \mid q \rangle$	$-q$	$\bar{p}$
$\langle p \mid \perp \mid q \rangle$	1	0
$\langle p \mid \top \mid q \rangle$	-1	0
$\langle p \mid \dashv \mid q \rangle$	0	1
$\langle p \mid \vdash \mid q \rangle$	0	-1
$\langle p \mid \hat{1} \mid q \rangle$	0	0
$\langle p \mid \hat{0} \mid q \rangle$	0	0

Table 2.6 The derivatives for the basic set of logic functions

Derivatives make it possible to detect a clear difference between the basic set of logic operators and the new compound operators of continuous logic. The derivatives for compound operators are constants:

α	$\frac{d\alpha}{dp}$	$\frac{d\alpha}{dq}$
$<\mid\vee + \wedge\mid>$	1	1
$<\mid\rightarrow + /\mid>$	-1	1
$<\mid\leftarrow + \backslash\mid>$	1	-1
$<\mid\downarrow + \uparrow\mid>$	-1	-1

The derivatives of logic functions can be employed to express a logic function. Consider, for instance, the disjunction function expressed via the derivatives of implication and converse implication:

$$<p\mid\vee\mid q> = \rightarrow_q' + \leftarrow_p' - \rightarrow_q'\leftarrow_p'$$

where for ease of notation we write $\rightarrow_q'$ in place of $<p\mid\rightarrow\mid q>_q'$.

Similarly, the NOR function is expressed in terms of the derivatives of disjunction as follows:

$$<p\mid\uparrow\mid q> \; = \; \vee_p' \vee_q' \; = \; <\overline{p\mid\wedge\mid q}>$$

<u>The graphs of logic derivatives</u>

The derivatives of tautology $\hat{1}(p,q)$ by either variable equals zero,

$$<p\mid\hat{1}\mid q>_p' \; = \; (\bar{p}\,\bar{q} + \bar{p}\,q + p\,\bar{q} + p\,q)_p' \; =$$

$$= \; -\,\bar{q} - q + \bar{q} + q \; = \; 0\,,$$

which might be evident without actual computation because tautology is constant:

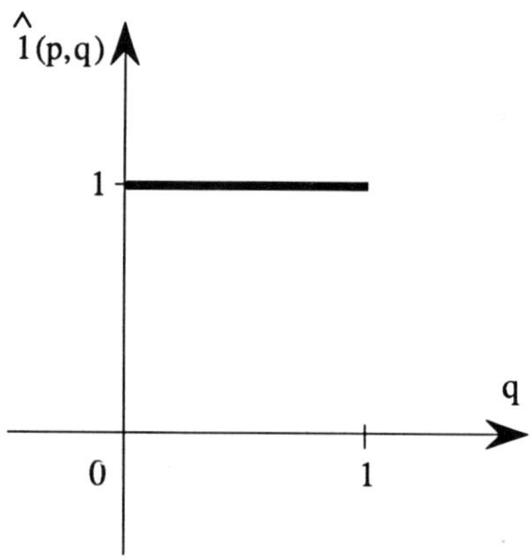

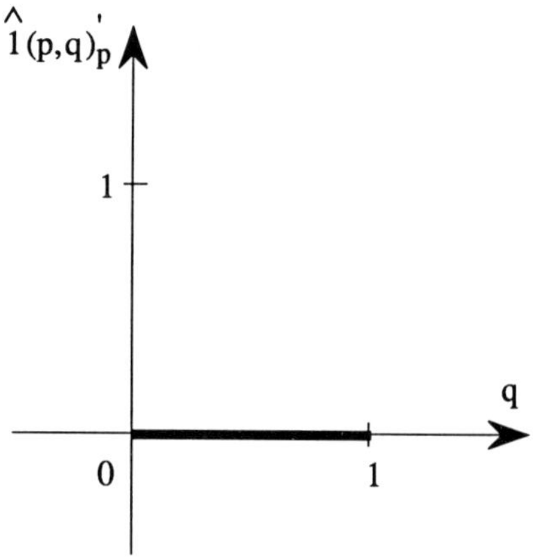

Because tautology is the sum of negation and affirmation

$$\hat{1} = \rceil + I,$$

it follows that

$$(\rceil + I)'_p = \rceil'_p + I'_p = 0$$

and therefore

$$\rceil'_p = -I'_p .$$

This is clearly seen when we plot the two derivatives together:

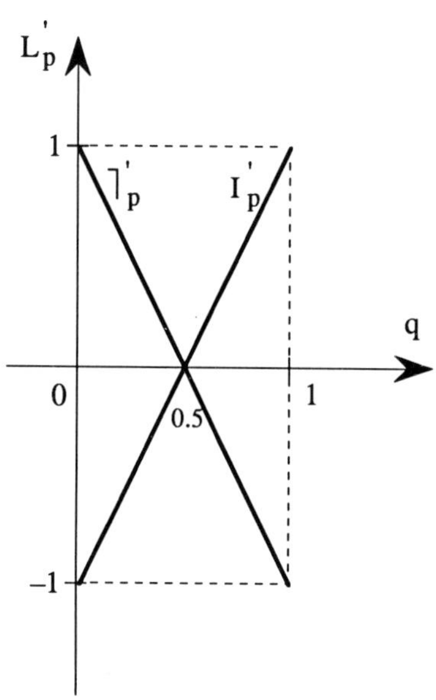

The importance of logic derivatives resides in the fact that the operations of differentiation and integration acquire the role of logic operators. This is most evident for the derivatives of tautology functions, where the effect of differentiation coincides with the effect of complementation:

$$<p|\,\hat{1}\,|q>'_p \;=\; \overline{<p|\,\hat{1}\,|q>}\,.$$

The result of both operations is a contradiction.

Derivatives of implication

For implication the derivative by consequent delivers the antecedent.

Implication:

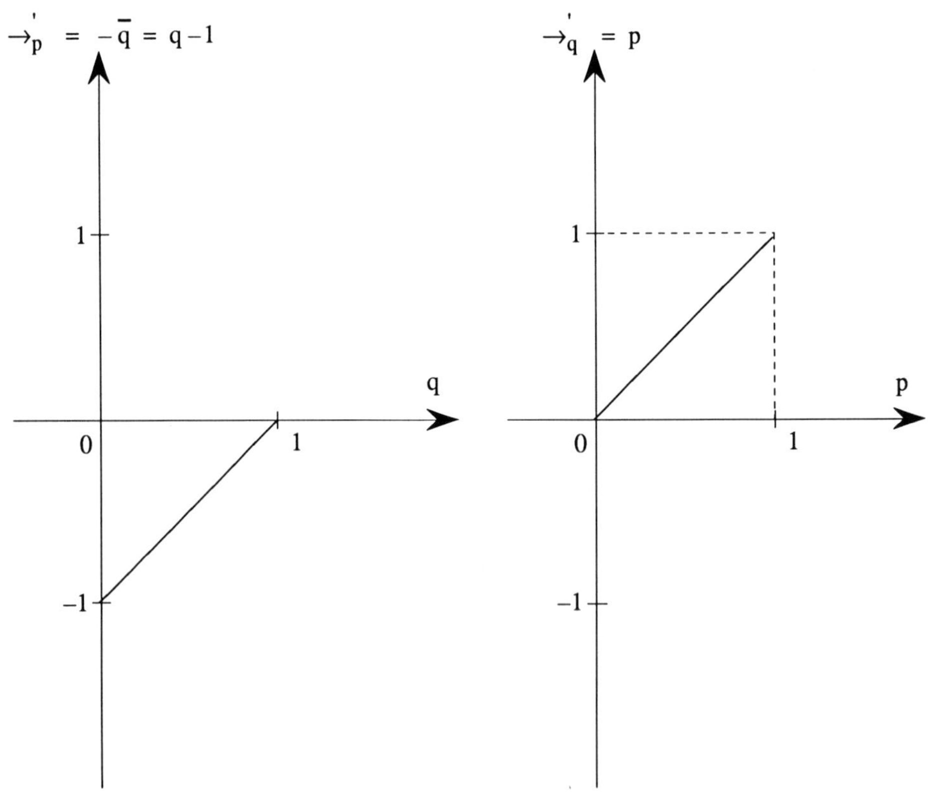

Converse implication:

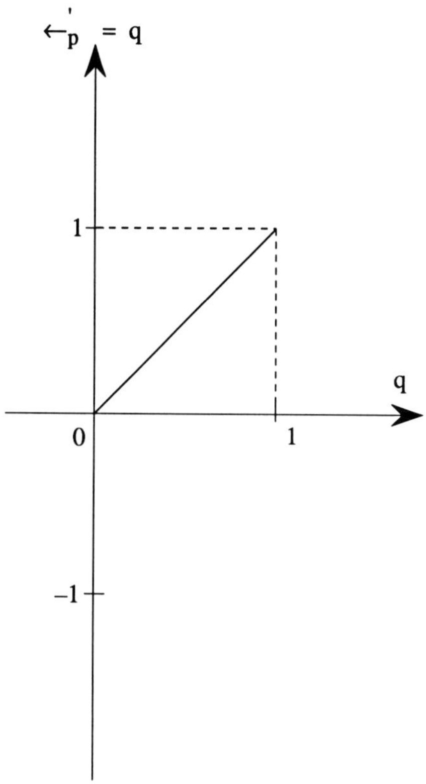

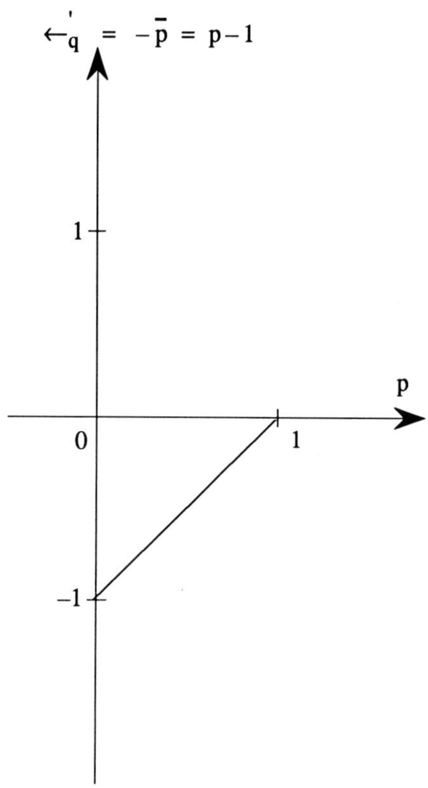

Because

$$\rightarrow'_p + 1 = (\theta\rightarrow)'_p$$

and

$$\rightarrow'_q = (\theta\rightarrow)'_q + 1\,,$$

where θ is the orthogonal transformation, we obtain differential tautologies:

$$(\theta\rightarrow)'_p - \rightarrow'_p = (\theta\rightarrow - \rightarrow)'_p \equiv 1$$

and

$$\to'_q - (\theta \to)'_q = (\to - \theta \to)'_q \equiv 1 .$$

Consequently,

$$<p \mid \theta \to - \to \mid q >'_p \equiv <p \mid \to - \theta \to \mid q >'_q .$$

Indeed,

$$<p \begin{pmatrix} 0 & -1 \\ 1 & 0 \end{pmatrix} q >'_p = <p \begin{pmatrix} 0 & 1 \\ -1 & 0 \end{pmatrix} q >'_q$$

$$\left(-\bar{p}\, q + p\, \bar{q}\right)'_p = \left(\bar{p}\, q - p\, \bar{q}\right)'_q$$

$$(p - q)'_p = (q - p)'_q$$

$$1 = 1$$

Differentiating implication with respect to antecedent and consequent and setting the results equal to zero, we obtain the logic condition under which the implication function $<p \mid \to \mid q>$ is at an extremum:

$$\to'_p = q - 1 = 0 \quad \Rightarrow q = 1$$

$$\to'_q = p = 0$$

This mathematical evaluation fits the logical situation precisely. It is exactly at these points, $p = 0$ and $q = 1$, that implication yields maximal truth-values:

$$\forall q : \quad <0 \mid \to \mid q> = 1$$

$$\forall p : \quad <p \mid \to \mid 1> = 1$$

We have linked the pure logical concept of validity to the mathematical concept of the extremum of a function, obtained by differentiation.

Constant derivatives

$$\bot'_p = 1 \qquad \top'_p = -1$$

$$\bot'_q = 0 \qquad \top'_q = 0$$

$$\dashv'_p = 0 \qquad \vdash'_p = 0$$

$$\dashv'_q = 1 \qquad \vdash'_q = -1$$

The derivatives of filters are constants:

$$\{1, -1, 0\} .$$

The negative result, –1, reflects the fact that the logic function is under negation, and the null result indicates that the filter was differentiated with respect to the variable on which it does not depend.

In the chart below we present jointly the graphs of the derivatives for the 16 basic logic functions:

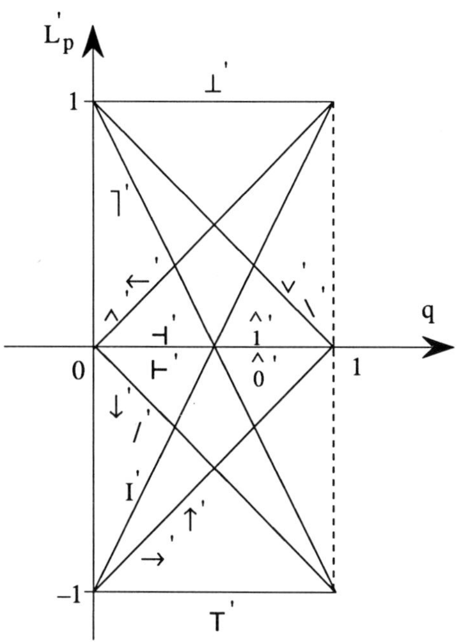

The derivative functions form a symmetric pattern where L' and $(\theta L)'$ lie on parallel lines:

$\{\wedge', \leftarrow'\}$	and	$\{\uparrow', \rightarrow'\}$
$\{\vee', \backslash'\}$	and	$\{\downarrow', /'\}$
$[\perp'\}$	and	$\{\top'\}$

or on the same line:

$$\dashv' \text{ and } \vdash', \qquad \hat{1}' \text{ and } \hat{0}'$$

and L' and (L*)', for $L^* \neq L \neq \bar{L}$, lie on orthogonal lines:

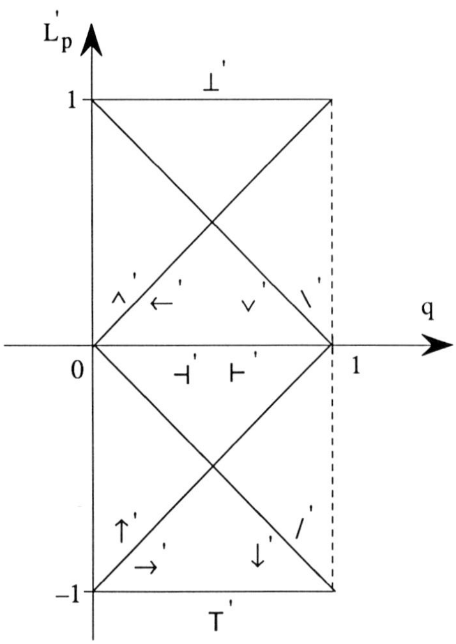

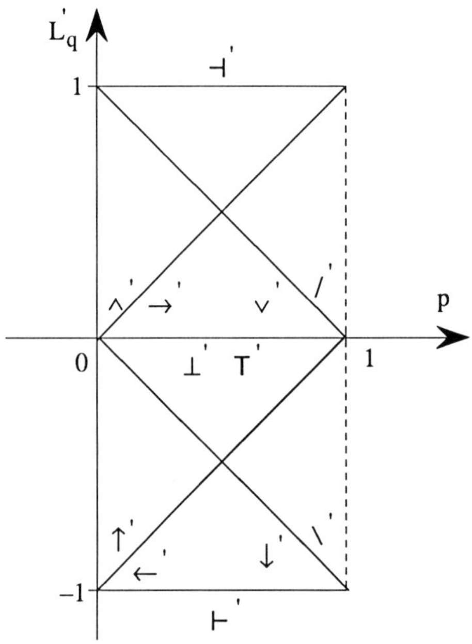

All second-order derivatives with respect to the same variable equal zero:

$$\frac{d^2 L(p,q)}{dp^2} = 0$$

$$\frac{d^2 L(p,q)}{dq^2} = 0$$

Second-order mixed derivatives form three sets:

for functions $\{\hat{1}, \hat{0}, \perp, \top, \dashv, \vdash\}$

$$\frac{d^2 L(p,q)}{dp\, dq} = 0 \, ;$$

for functions $\{\wedge, \backslash, /, \uparrow, \vee, \rightarrow, \leftarrow, \downarrow\}$

$$\frac{d^2 L(p,q)}{dp\, dq} = \pm 1 \, ;$$

for functions $\{I, \urcorner\}$

$$\frac{d^2 L(p,q)}{dp\, dq} = \pm 2 \, .$$

The second-order mixed derivatives commute:

$$\frac{d^2 L(p,q)}{dp\, dq} = \frac{d^2 L(p,q)}{dq\, dp} \, .$$

Logic integration constants

We have formulated the rule for inductive inference as the subtraction of contradiction, that is, subtraction or addition of 0. In the context of differential logic, inductive inference can be viewed as the result of the differentiation of a logic function that contains a constant. Since the integral of a function is defined up to a constant, this provides a natural linkage of the inductive operations with the boundary conditions problem. A simple rule for computing the logic integration constants can be established for the basic cases:

α	$\frac{d\alpha}{dp}$	C_o	$\alpha = \int \frac{d\alpha}{dp}\,dp + C_o$
$< p \mid \perp \mid q > \; = p$	1	0	$\int_0^p 1\,dp + 0 = p$
$< p \mid T \mid q > \; = \bar{p}$	-1	1	$\int_0^p -1\,dp + 1 = 1 - p = \bar{p}$
$< p \mid \wedge \mid q > \; = p\,q$	q	0	$\int_0^p q\,dp + 0 = p\,q$
$< p \mid \backslash \mid q > \; = p\,\bar{q}$	$\bar{q}$	0	$\int_0^p \bar{q}\,dp + 0 = p\,\bar{q}$
$< p \mid / \mid q > \; = \bar{p}\,q$	$-q$	q	$\int_0^p -q\,dp + q = q - p\,q = \bar{p}\,q$
$< p \mid \uparrow \mid q > \; = \bar{p}\,\bar{q}$	$-\bar{q}$	$\bar{q}$	$\int_0^p -\bar{q}\,dp + \bar{q} = \bar{q} - \bar{q}\,p = \bar{p}\,\bar{q}$

Table 2.7 Logic integration constants

As is indicated in the chart for positive-valued logic derivatives the integration constant is 0. For negative-valued derivatives the integration constant is the same function but with the opposite sign.

Differential duality laws

Under the differentiation operator the complement operator and the negative sign annihilate each other:

$$<p|L|q>' = \left(-<\overline{p|L|q}>\right)' .$$

This establishes a general rule for the elimination of the negative sign:

$$-<p|L|q>' = <\overline{p|L|q}>' = <p|\overline{L}|q>' .$$

We cancel minus but introduce the complement operator, which can be applied to an operator directly. Conversely, we can cancel the complement operator, simultaneously changing the sign to the opposite. Consequently, any duality relation in matrix logic,

$$<p|L|q> = \overline{<\overline{p}|L^*|\overline{q}>} ,$$

can be written in differential form as follows

$$<p|L|q>' = -<\overline{p}|L^*|\overline{q}>' .$$

For the duality transformations of conjunction and disjunction we obtain:

$$\begin{array}{ccc} <p|\wedge|q>'_p & = & -<\overline{p}|\vee|\overline{q}>'_p \\ q & = & -(-q) \end{array}$$

and

$$\begin{array}{ccc} <p|\vee|q>'_p & = & -<\overline{p}|\wedge|\overline{q}>'_p \\ \overline{q} & = & -(-\overline{q}) \end{array}$$

Among the various dual operators conjunction and disjunction are distinguished by also having special differential pseudoduality relations:

$$\begin{array}{ccc} <p|\wedge|q>'_p & = & \overline{<p|\vee|q>'_p} \\ q & = & \overline{\overline{q}} \end{array}$$

and

$$\begin{array}{ccc} <p|\vee|q>'_p & = & \overline{<p|\wedge|q>'_p} \\ \overline{q} & = & \overline{q} \end{array}$$

Note that in pseudoduality transformations the complement operator operates over the differentiation operator, whereas in duality transformations it operates under the differentiation operator.

The special pseudoduality trnasformations hold exclusively for the conjunction and disjunction functions, clearly because they are the only duals whose derivatives are complementary.

The differential pseudoduality laws for conjunction and disjunction can be reformulated as follows:

$$
\begin{aligned}
\overline{\langle p | \wedge | q \rangle'} &= \langle p | \vee | q \rangle' = \\
&= \langle p | \overline{\theta \wedge} | q \rangle' = \\
&= \langle \overline{p | \theta \wedge | q} \rangle' = \\
&= \langle \overline{\overline{p} | \wedge | \overline{q}} \rangle' ,
\end{aligned}
$$

and

$$
\begin{aligned}
\overline{\langle p | \vee | q \rangle'} &= \langle p | \wedge | q \rangle' = \\
&= \langle p | \overline{\theta \vee} | q \rangle' = \\
&= \langle \overline{p | \theta \vee | q} \rangle' = \\
&= \langle \overline{\overline{p} | \vee | \overline{q}} \rangle' .
\end{aligned}
$$

For the fundamental logic operators of conjunction and disjunction the order of application of the complement and differentiation operators can be reversed if the logic vectors in the bracket product are negated. In other words, to obtain commutation for the operator of differentiation and the complement operator, the logic vectors must be complemented.

Extremums

By differentiating logic functions and setting the results to zero, we obtain a condition for an extremum for logic functions. Extremums reflect important logic properties. It follows, as expected, that the implication functions have an extremum with the consequent equal to 1 or the antecedent equal to 0.

Verum ex quolibet:

$$\to'_p = -\bar{q} = q - 1 = 0$$

$$q = 1$$

Ex falso quodlibet:

$$\to'_q = p = 0$$

Less evident is the presence of a saddle point for the affirmation and negation functions,

$$\begin{cases} I'_p = 2q - 1 = 0\,, & q = \frac{1}{2} \\ I'_q = 2p - 1 = 0\,, & p = \frac{1}{2} \end{cases}$$

$$\begin{cases} \urcorner'_p = 1 - 2q = 0\,, & q = \frac{1}{2} \\ \urcorner'_q = 1 - 2p = 0\,, & p = \frac{1}{2} \end{cases}$$

For both functions the saddle point occurs at $p = q = \frac{1}{2}$. The minimum of affirmation equals the maximum of negation and vice versa.

The important problem of the assignment of fractional truth-values which evaluate a compound logic function to a maximum can be solved with the methods of differential logic which allow the formulation of logic problems as problems of optimization and linear programming. The concept of logic derivatives opens a vast domain of unexplored possibilities, and paves the way for incorporating logic fully into the framework of the exact sciences.

Intelligence code

With the extension of logic into the continuous universe we can now provide an exact mathematical formulation of the intelligence code, which as a law of conservation takes the form

$$\sum^{i} \frac{\partial L(q_i)}{\partial q_i} = 0 \qquad i = 1, \ldots, 2n$$

Here the differentiation operator acts upon an inference rule expressed as the logic function $L(q_i)$ of n variables, where falsum and verum components are treated as Hamiltonian-type canonical pairs of coordinates in logic space. Because any inference rule is a tautology for a logic function of two variables, the conservation law is formulated as follows:

$$\frac{\partial L(p,q)}{\partial p} + \frac{\partial L(p,q)}{\partial \overline{p}} = 0$$

$$\frac{\partial L(p,q)}{\partial q} + \frac{\partial L(p,q)}{\partial \overline{q}} = 0$$

which implies

$$\frac{\partial L(p,q)}{\partial p} = - \frac{\partial L(p,q)}{\partial \overline{p}}$$

$$\frac{\partial L(p,q)}{\partial q} = - \frac{\partial L(p,q)}{\partial \overline{q}}$$

When it is assumed that logical truth is established directly, it is, in fact, established through the elimination of falsity. The covariant operations of cognition require that verum and falsum potentials balance each other, analogous to the relation between kinetic and potential energy. In common terms, the acquisition of knowledge is recorded through the dissipation of ignorance and vice versa. It follows that the logic process remains invariant whenever a transition from verum to falsum is accompanied by complementation of a logic operator and vice versa:

$$\sum_{i=1}^{n} \frac{\partial L(q_i)}{\partial q_i} = \sum_{i=1}^{n} \frac{\partial \overline{L}(q_i)}{\partial \overline{q}_i}$$

The symmetry in the properties of operators appears as conservation laws in logic interactions. The realization that the laws of conservation provide a foundation on which the intelligence code is based forms a new and powerful paradigm for the study of the logic mechanisms of the mind.

CHAPTER 3

INTELLIGENT INTERACTIONS

The conclusion that logic should model itself on the exact sciences calls for a new and challenging programme for a reform of logic theory. As a matter of fact, logic throughout its history has always followed in the footsteps of mathematics and physics. And just as conventional logic has bridged logic and classical physics, matrix logic can serve as a unifying link between logic and fundamental physics.

3.1 Antilogic

Conventional logic algebra is based on the idempotency axiom

$$x^2 = x ,$$

from which the fundamental contradiction law derives naturally. Because the idempotency axiom is satisfied only by the positive integers 0 and 1, standing for false and true respectively, the conclusion is reached that the logic division of the mind is twofold. Accepting symmetries of a higher order, in matrix logic we postulate the prime number axiom,

$$x^{\alpha} = x , \qquad \alpha = 1, 2, 3, 5, \ldots$$

For $\alpha = 3$ we obtain the cubic equation

$$x^3 = x ,$$

which, in addition to 0 and 1, is also satisfied by the negative value –1! For conventional logic this fact is logically uninterpretable. But in matrix logic it allows one to incorporate into logic analysis mental divisions of a higher degree and thus to achieve the fundamental extension of logic into the negative domain.

The matrix operator formulation of logic provides compelling reasons for the development of a negative logic Ω^-. As a matter of fact, the need to consider the extension of logic into the negative domain is dictated by the matrix logic formalism itself.

3.1.1 Matrix inverses

In classical logic the negation of negation is identity:

$$\overline{\overline{x}} = x .$$

This postulate implicitly entails that the negation connective is the inverse of itself and, being applied iteratively, cancels itself out. Negation is the only inverse operation employed in conventional logic. In matrix logic this axiomatic definition of the inverse logic operation is replaced by the rigour of mathematical principle: the square of the negation operator is a unit matrix:

$$\rceil^2 = I ,$$

which implies that the negation operator is a matrix inverse of itself:

$$\rceil = \rceil^{-1}$$

and therefore

$$\rceil\rceil = \rceil\rceil^{-1} = \rceil^{-1}\rceil = I .$$

Since the notion of inverse obtains in matrix logic the constructive meaning of an inverse matrix, this suggests the idea of investigating in this framework the other logic operators from the set Ω+. The table on the next page provides the complete set of logic inverses. Of the six non-singular matrix operators, two, equivalence and negation, are inverses of themselves, and, consequently, their inverses belong to the same set Ω+. But the four remaining inverses, those of disjunction, implication, converse implication and NAND

$$\vee^{-1}, \rightarrow^{-1}, \leftarrow^{-1}, \downarrow^{-1}$$

are operators of a new kind, containing the negative sign for one of the matrix elements.
Till now all formalizations of logic have employed positive number representations. An attempt to interpret consistently negative values of the matrix elements of logic inverses leads to a fundamental extension of the domain of logic. In addition to positive matrix logic, which we described in the second chapter, a negative logic, with very interesting properties, can be developed. The design of negative logic, as the following material will show, reveals new and important facets of logic.

L	det L	L^{-1}
$\leftrightarrow$	1	$\begin{pmatrix} 1 & 0 \\ 0 & 1 \end{pmatrix}$
$\neg$	− 1	$\begin{pmatrix} 0 & 1 \\ 1 & 0 \end{pmatrix}$
$\wedge$	0	—
$\vee$	− 1	$\begin{pmatrix} -1 & 1 \\ 1 & 0 \end{pmatrix}$
$\rightarrow$	1	$\begin{pmatrix} 1 & -1 \\ 0 & 1 \end{pmatrix}$
$\leftarrow$	1	$\begin{pmatrix} 1 & 0 \\ -1 & 1 \end{pmatrix}$
$\backslash$	0	—
$/$	0	—
$\uparrow$	0	—
$\downarrow$	− 1	$\begin{pmatrix} 0 & 1 \\ 1 & -1 \end{pmatrix}$
$\perp$	0	—
$\top$	0	—
$\dashv$	0	—
$\vdash$	0	—
$\hat{1}$	0	—
$\hat{0}$	0	—

Table 3.1 Inverses of logic operators

In order to distinguish between the set of operators of positive logic Ω^+ and the set of operators of negative logic Ω^- we assign corresponding superscripts. Similarly, L^+ will indicate an operator from the set Ω^+, and L^- an operator from the set Ω^-.
Considering all possible distributions of the negative unit we obtain the complete set of matrix logic operators including its positive and negative subsets:

	Ω^+	Ω^-
Conjunction	$\begin{pmatrix} 0 & 0 \\ 0 & 1 \end{pmatrix}$	$\begin{pmatrix} 0 & 0 \\ 0 & -1 \end{pmatrix}$
Inhibition	$\begin{pmatrix} 0 & 0 \\ 1 & 0 \end{pmatrix}$	$\begin{pmatrix} 0 & 0 \\ -1 & 0 \end{pmatrix}$
Converse Inhibition	$\begin{pmatrix} 0 & 1 \\ 0 & 0 \end{pmatrix}$	$\begin{pmatrix} 0 & -1 \\ 0 & 0 \end{pmatrix}$
Nor	$\begin{pmatrix} 1 & 0 \\ 0 & 0 \end{pmatrix}$	$\begin{pmatrix} -1 & 0 \\ 0 & 0 \end{pmatrix}$
Equivalence	$\begin{pmatrix} 1 & 0 \\ 0 & 1 \end{pmatrix}$	$\begin{pmatrix} -1 & 0 \\ 0 & 1 \end{pmatrix} \begin{pmatrix} 1 & 0 \\ 0 & -1 \end{pmatrix} \begin{pmatrix} -1 & 0 \\ 0 & -1 \end{pmatrix}$
Negation	$\begin{pmatrix} 0 & 1 \\ 1 & 0 \end{pmatrix}$	$\begin{pmatrix} 0 & -1 \\ 1 & 0 \end{pmatrix} \begin{pmatrix} 0 & 1 \\ -1 & 0 \end{pmatrix} \begin{pmatrix} 0 & -1 \\ -1 & 0 \end{pmatrix}$
Filter	$\begin{pmatrix} 0 & 0 \\ 1 & 1 \end{pmatrix}$	$\begin{pmatrix} 0 & 0 \\ -1 & 1 \end{pmatrix} \begin{pmatrix} 0 & 0 \\ 1 & -1 \end{pmatrix} \begin{pmatrix} 0 & 0 \\ -1 & -1 \end{pmatrix}$
Antifilter	$\begin{pmatrix} 1 & 1 \\ 0 & 0 \end{pmatrix}$	$\begin{pmatrix} -1 & 1 \\ 0 & 0 \end{pmatrix} \begin{pmatrix} 1 & -1 \\ 0 & 0 \end{pmatrix} \begin{pmatrix} -1 & -1 \\ 0 & 0 \end{pmatrix}$

Converse filter	$\begin{pmatrix} 0 & 1 \\ 0 & 1 \end{pmatrix}$	$\begin{pmatrix} 0 & -1 \\ 0 & 1 \end{pmatrix}\begin{pmatrix} 0 & 1 \\ 0 & -1 \end{pmatrix}\begin{pmatrix} 0 & -1 \\ 0 & -1 \end{pmatrix}$
Converse antifilter	$\begin{pmatrix} 1 & 0 \\ 1 & 0 \end{pmatrix}$	$\begin{pmatrix} -1 & 0 \\ 1 & 0 \end{pmatrix}\begin{pmatrix} 1 & 0 \\ -1 & 0 \end{pmatrix}\begin{pmatrix} -1 & 0 \\ -1 & 0 \end{pmatrix}$
Disjunction	$\begin{pmatrix} 0 & 1 \\ 1 & 1 \end{pmatrix}$	$\begin{pmatrix} 0 & 1 \\ 1 & -1 \end{pmatrix}\begin{pmatrix} 0 & 1 \\ -1 & 1 \end{pmatrix}\begin{pmatrix} 0 & -1 \\ 1 & 1 \end{pmatrix}\begin{pmatrix} 0 & 1 \\ -1 & -1 \end{pmatrix}\begin{pmatrix} 0 & -1 \\ 1 & -1 \end{pmatrix}\begin{pmatrix} 0 & -1 \\ -1 & 1 \end{pmatrix}\begin{pmatrix} 0 & -1 \\ -1 & -1 \end{pmatrix}$
Nand	$\begin{pmatrix} 1 & 1 \\ 1 & 0 \end{pmatrix}$	$\begin{pmatrix} -1 & 1 \\ 1 & 0 \end{pmatrix}\begin{pmatrix} 1 & -1 \\ 1 & 0 \end{pmatrix}\begin{pmatrix} 1 & 1 \\ -1 & 0 \end{pmatrix}\begin{pmatrix} -1 & -1 \\ 1 & 0 \end{pmatrix}\begin{pmatrix} -1 & 1 \\ -1 & 0 \end{pmatrix}\begin{pmatrix} 1 & -1 \\ -1 & 0 \end{pmatrix}\begin{pmatrix} -1 & -1 \\ -1 & 0 \end{pmatrix}$
Implication	$\begin{pmatrix} 1 & 1 \\ 0 & 1 \end{pmatrix}$	$\begin{pmatrix} 1 & -1 \\ 0 & 1 \end{pmatrix}\begin{pmatrix} 1 & 1 \\ 0 & -1 \end{pmatrix}\begin{pmatrix} -1 & 1 \\ 0 & 1 \end{pmatrix}\begin{pmatrix} -1 & -1 \\ 0 & 1 \end{pmatrix}\begin{pmatrix} 1 & -1 \\ 0 & -1 \end{pmatrix}\begin{pmatrix} -1 & 1 \\ 0 & -1 \end{pmatrix}\begin{pmatrix} -1 & -1 \\ 0 & -1 \end{pmatrix}$
Converse Implication	$\begin{pmatrix} 1 & 0 \\ 1 & 1 \end{pmatrix}$	$\begin{pmatrix} 1 & 0 \\ -1 & 1 \end{pmatrix}\begin{pmatrix} -1 & 0 \\ 1 & 1 \end{pmatrix}\begin{pmatrix} 1 & 0 \\ 1 & -1 \end{pmatrix}\begin{pmatrix} 1 & 0 \\ -1 & -1 \end{pmatrix}\begin{pmatrix} -1 & 0 \\ -1 & 1 \end{pmatrix}\begin{pmatrix} -1 & 0 \\ 1 & -1 \end{pmatrix}\begin{pmatrix} -1 & 0 \\ -1 & -1 \end{pmatrix}$
Tautology	$\begin{pmatrix} 1 & 1 \\ 1 & 1 \end{pmatrix}$	$\begin{pmatrix} -1 & 1 \\ 1 & 1 \end{pmatrix}\begin{pmatrix} 1 & 1 \\ -1 & 1 \end{pmatrix}\begin{pmatrix} 1 & -1 \\ 1 & 1 \end{pmatrix}\begin{pmatrix} 1 & 1 \\ 1 & -1 \end{pmatrix}$ $\begin{pmatrix} -1 & 1 \\ 1 & -1 \end{pmatrix}\begin{pmatrix} 1 & -1 \\ -1 & 1 \end{pmatrix}\begin{pmatrix} 1 & 1 \\ -1 & -1 \end{pmatrix}\begin{pmatrix} -1 & -1 \\ 1 & 1 \end{pmatrix}\begin{pmatrix} 1 & -1 \\ 1 & -1 \end{pmatrix}\begin{pmatrix} -1 & 1 \\ -1 & 1 \end{pmatrix}$ $\begin{pmatrix} 1 & -1 \\ -1 & -1 \end{pmatrix}\begin{pmatrix} -1 & -1 \\ -1 & 1 \end{pmatrix}\begin{pmatrix} -1 & 1 \\ -1 & -1 \end{pmatrix}\begin{pmatrix} -1 & -1 \\ 1 & -1 \end{pmatrix}$ $\begin{pmatrix} -1 & -1 \\ -1 & -1 \end{pmatrix}$
Empty operator	$\begin{pmatrix} 0 & 0 \\ 0 & 0 \end{pmatrix}$	$\begin{pmatrix} 0 & 0 \\ 0 & 0 \end{pmatrix}$

Table 3.2 Complete set of logic operators

With each operator of positive logic L_i^+, i = 1, 16, we associated a corresponding series of negative logic operators, with a different distribution of negative units within the logic operators but which by modulo are all equal. Thus the negative inverses which were derived in the preceding section are just a few negative logic operators from the much larger set Ω^-.

While the domain of positive logic functions can be conveniently described by the unit square, we must represent the complete domain of matrix logic, now taking into account, in addition to 0 and 1, the truth antivalue –1, by two unit domains that are opposite in value:

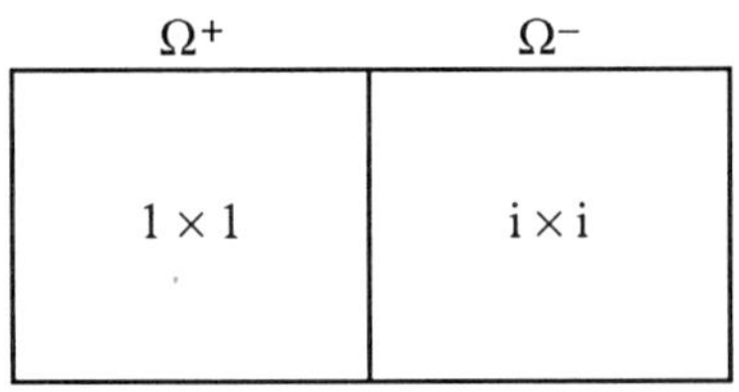

Here the real square $1 \times 1 = 1$ corresponds to positive logic Ω^+, and the imaginary square $i \times i = -1$ to negative logic Ω^-. This clearly distinguishes matrix logic algebra from probability algebra, which is defined over the unit universe 1×1. Whereas a probability algebra can be built over a set of numbers whose sum is 1, matrix logic algebra is built over a set of numbers whose sum is zero, that is, whose modulo sum is 2.
Introduction of the negative truth value –1 obliges us to redefine the normalization condition for logic vectors, which now takes the form

$$|\bar{p}| + |p| = 1 ,$$

where the verum and falsum components are taken by modulo.
With rotation of logic coordinates, logic vectors that are logically normalized run along the sides of the characteristic square where the normalization condition is satisfied. Inside and outside the characteristic square lie two areas of virtual logic states:

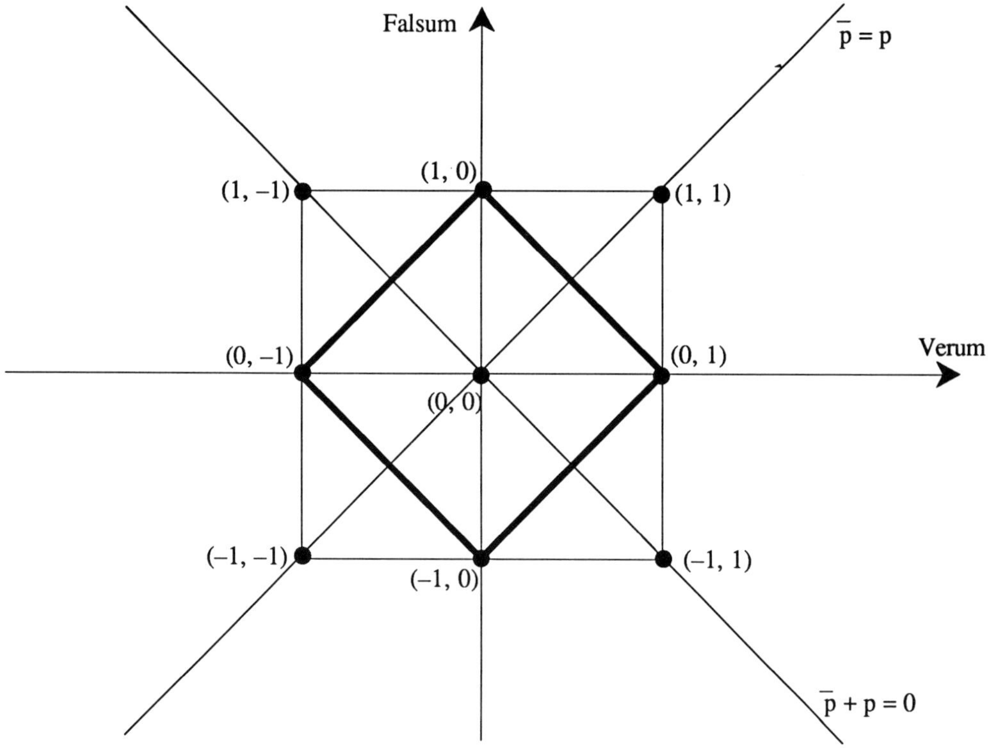

A grid of 9 points, indicated in the graph, defines the domain of discrete matrix logic; this compares with the Boolean lattice of just two points:

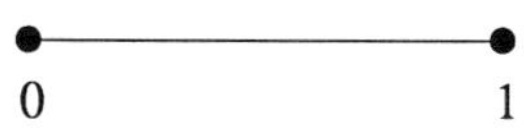

Whereas in Ω^{+} logic the values of the logic functions are defined in the binary set

$$E_2 = \{0,1\} ,$$

in Ω^{-} logic, with the introduction of the negative value, the logic functions are defined in the ternary balanced set

$$E_3 = \{-1, 0, 1\} .$$

with the special strictly negative subset of 16 operators defined in the negative binary set

$$E_2^{(-)} = \{-1, 0\} .$$

The operators of negative logic and the operators of positive logic act identically in the spaces of strictly negative and strictly positive logic vectors:

$$<x \mid L^{+} \mid y> = <-x \mid L^{+} \mid -y> ,$$

$$<x \mid L^{-} \mid y> = <-x \mid L^{-} \mid -y> .$$

In this case negative logic remains negative and positive logic remains positive, with the outcome being dependent not on the operators used but on the spaces in which they act.
However, if we consider the cross-interaction of positive and negative logic spaces where the logic operator acts between negative and positive logics the situation becomes more complex. A positive logic operator L^+ acting between these two logics will produce one of the negative logic functions, defined on the set $E_2^{(-)}$:

$$<x \mid L^{+} \mid -y> = <x \mid L^{-} \mid y> ,$$

with

$$L^{-} = - L^{+} .$$

But application of a negative logic operator L^- in the same circumstances can result either in a positive logic function or in a negative logic function, depending which operator from the set Ω^- is used:

$$<x \mid L^{-} \mid -y> = \begin{cases} <x \mid L_i^{+} \mid y> , & i = 1, 16 . \\ <x \mid L_j^{-} \mid y> , & j = 1, 50 . \end{cases}$$

The asymmetry of the relationships between positive and negative logic is apparent in fundamental products, where a product containing or even consisting entirely of negative logic functions actually represents a positive logic function:

$$<x \mid \vee \mid y><x \mid \vee^{-1} \mid y> \;=\; <\overline{x \mid y}>$$

$$<x \mid \rightarrow^{-1} \mid y><x \mid \leftarrow^{-1} \mid y> \;=\; <x \mid y>$$

Negative-valued operators can also appear in the fundamental products of positive matrix logic, though the final results of evaluation belong exclusively to the binary set $E_2 = \{0,1\}$.

Because the fundamental product is an algebraic sum

$$\alpha = \sum \pm \wedge \langle x_i \mid L_j \mid y_k \rangle ,$$

for those summands which enter into a product with a negative sign, the latter can always be transferred onto an operator

$$- \langle x \mid L \mid y \rangle = \langle x \mid -L \mid y \rangle ,$$

which could lead to the formation of 15 pure negative operators only. It is necessary to emphasize that the negative component of logic is here entirely hidden, and can never show itself in the final results of the evaluation of a fundamental product. This hidden property is an indication that the mind can operate equally well either in pure positive logic, $E_2 = \{0,1\}$, or pure negative logic, $E_2^{(-)} = \{-1, 0\}$, apparently being unable to detect the difference, since nothing would indicate it. Negative logic begins to manifest itself only through interaction with positive logic.

In the psychological framework positive and negative logics can be associated with conscious and unconscious operations respectively. In view of the fact that a major part of intelligent activity is generally hidden from conscious control, the role of negative logic in the operations of the mind must be significant.

Multiplication table of negative logic

Because in negative logic we operate in the three-valued set E_3 instead of the familiar binary set E_2, we encounter symmetries of a more complex and richer nature than those of positive logic alone. Whereas the alphabet of truth-values in negative logic increases, by comparison with positive logic, from 2 to 3, the number of logic functions grows from 16 in positive logic to 81 in $\Omega^+ \cap \Omega^-$.

The presentation of the multiplication table of negative logic is, for obvious reasons, a significantly larger task than the derivation of the multiplication table for the set Ω^+. By way of illustration we consider the multiplication and transmultiplication tables for the implication series:

$\times$	$\begin{pmatrix}1&1\\0&1\end{pmatrix}$	$\begin{pmatrix}-1&1\\0&1\end{pmatrix}$	$\begin{pmatrix}1&-1\\0&1\end{pmatrix}$	$\begin{pmatrix}1&1\\0&-1\end{pmatrix}$	$\begin{pmatrix}-1&-1\\0&1\end{pmatrix}$	$\begin{pmatrix}1&-1\\0&-1\end{pmatrix}$	$\begin{pmatrix}-1&1\\0&-1\end{pmatrix}$	$\begin{pmatrix}-1&-1\\0&-1\end{pmatrix}$
$\begin{pmatrix}1&1\\0&1\end{pmatrix}$			$\begin{pmatrix}1&0\\0&1\end{pmatrix}$	$\begin{pmatrix}1&0\\0&-1\end{pmatrix}$	$\begin{pmatrix}-1&0\\0&1\end{pmatrix}$		$\begin{pmatrix}-1&0\\0&-1\end{pmatrix}$	
$\begin{pmatrix}-1&1\\0&1\end{pmatrix}$	$\begin{pmatrix}-1&0\\0&1\end{pmatrix}$	$\begin{pmatrix}1&0\\0&1\end{pmatrix}$				$\begin{pmatrix}-1&0\\0&-1\end{pmatrix}$		$\begin{pmatrix}1&0\\0&-1\end{pmatrix}$
$\begin{pmatrix}1&-1\\0&1\end{pmatrix}$	$\begin{pmatrix}1&0\\0&1\end{pmatrix}$	$\begin{pmatrix}-1&0\\0&1\end{pmatrix}$				$\begin{pmatrix}1&0\\0&-1\end{pmatrix}$		$\begin{pmatrix}-1&0\\0&-1\end{pmatrix}$
$\begin{pmatrix}1&1\\0&-1\end{pmatrix}$			$\begin{pmatrix}1&0\\0&-1\end{pmatrix}$	$\begin{pmatrix}1&0\\0&1\end{pmatrix}$	$\begin{pmatrix}-1&0\\0&-1\end{pmatrix}$		$\begin{pmatrix}-1&0\\0&1\end{pmatrix}$	
$\begin{pmatrix}-1&-1\\0&1\end{pmatrix}$			$\begin{pmatrix}-1&0\\0&1\end{pmatrix}$	$\begin{pmatrix}-1&0\\0&-1\end{pmatrix}$	$\begin{pmatrix}1&0\\0&1\end{pmatrix}$		$\begin{pmatrix}1&0\\0&-1\end{pmatrix}$	
$\begin{pmatrix}1&-1\\0&-1\end{pmatrix}$	$\begin{pmatrix}1&0\\0&-1\end{pmatrix}$	$\begin{pmatrix}-1&0\\0&-1\end{pmatrix}$				$\begin{pmatrix}1&0\\0&1\end{pmatrix}$		$\begin{pmatrix}-1&0\\0&1\end{pmatrix}$
$\begin{pmatrix}-1&1\\0&-1\end{pmatrix}$	$\begin{pmatrix}-1&0\\0&-1\end{pmatrix}$	$\begin{pmatrix}1&0\\0&-1\end{pmatrix}$				$\begin{pmatrix}-1&0\\0&1\end{pmatrix}$		$\begin{pmatrix}1&0\\0&1\end{pmatrix}$
$\begin{pmatrix}-1&-1\\0&-1\end{pmatrix}$			$\begin{pmatrix}-1&0\\0&-1\end{pmatrix}$	$\begin{pmatrix}-1&0\\0&1\end{pmatrix}$	$\begin{pmatrix}1&0\\0&-1\end{pmatrix}$		$\begin{pmatrix}1&0\\0&1\end{pmatrix}$	

Table 3.3 Multiplication table for implication series

✳	$\begin{pmatrix}1&1\\0&1\end{pmatrix}$	$\begin{pmatrix}-1&1\\0&1\end{pmatrix}$	$\begin{pmatrix}1&-1\\0&1\end{pmatrix}$	$\begin{pmatrix}1&1\\0&-1\end{pmatrix}$	$\begin{pmatrix}-1&-1\\0&1\end{pmatrix}$	$\begin{pmatrix}1&-1\\0&-1\end{pmatrix}$	$\begin{pmatrix}-1&1\\0&-1\end{pmatrix}$	$\begin{pmatrix}-1&-1\\0&-1\end{pmatrix}$
$\begin{pmatrix}1&1\\0&1\end{pmatrix}$	$\rightarrow$	$\begin{pmatrix}-1&1\\0&1\end{pmatrix}$	$\begin{pmatrix}1&-1\\0&1\end{pmatrix}$	$\begin{pmatrix}1&1\\0&-1\end{pmatrix}$	$\begin{pmatrix}-1&-1\\0&1\end{pmatrix}$	$\begin{pmatrix}1&-1\\0&-1\end{pmatrix}$	$\begin{pmatrix}-1&1\\0&-1\end{pmatrix}$	$\begin{pmatrix}-1&-1\\0&-1\end{pmatrix}$
$\begin{pmatrix}-1&1\\0&1\end{pmatrix}$	$\begin{pmatrix}-1&1\\0&1\end{pmatrix}$	$\rightarrow$	$\begin{pmatrix}-1&-1\\0&1\end{pmatrix}$	$\begin{pmatrix}-1&1\\0&-1\end{pmatrix}$	$\begin{pmatrix}1&-1\\0&1\end{pmatrix}$	$\begin{pmatrix}-1&-1\\0&-1\end{pmatrix}$	$\begin{pmatrix}1&1\\0&-1\end{pmatrix}$	$\begin{pmatrix}1&-1\\0&-1\end{pmatrix}$
$\begin{pmatrix}1&-1\\0&1\end{pmatrix}$	$\begin{pmatrix}1&-1\\0&1\end{pmatrix}$	$\begin{pmatrix}-1&-1\\0&1\end{pmatrix}$	$\rightarrow$	$\begin{pmatrix}1&-1\\0&-1\end{pmatrix}$	$\begin{pmatrix}-1&1\\0&1\end{pmatrix}$	$\begin{pmatrix}1&1\\0&-1\end{pmatrix}$	$\begin{pmatrix}-1&-1\\0&-1\end{pmatrix}$	$\begin{pmatrix}-1&1\\0&-1\end{pmatrix}$
$\begin{pmatrix}1&1\\0&-1\end{pmatrix}$	$\begin{pmatrix}1&1\\0&-1\end{pmatrix}$	$\begin{pmatrix}-1&1\\0&-1\end{pmatrix}$	$\begin{pmatrix}1&-1\\0&-1\end{pmatrix}$	$\rightarrow$	$\begin{pmatrix}-1&-1\\0&-1\end{pmatrix}$	$\begin{pmatrix}1&-1\\0&1\end{pmatrix}$	$\begin{pmatrix}-1&1\\0&1\end{pmatrix}$	$\begin{pmatrix}-1&-1\\0&1\end{pmatrix}$
$\begin{pmatrix}-1&-1\\0&1\end{pmatrix}$	$\begin{pmatrix}-1&-1\\0&1\end{pmatrix}$	$\begin{pmatrix}1&-1\\0&1\end{pmatrix}$	$\begin{pmatrix}-1&1\\0&1\end{pmatrix}$	$\begin{pmatrix}-1&-1\\0&-1\end{pmatrix}$	$\rightarrow$	$\begin{pmatrix}-1&1\\0&-1\end{pmatrix}$	$\begin{pmatrix}1&-1\\0&-1\end{pmatrix}$	$\begin{pmatrix}1&1\\0&-1\end{pmatrix}$
$\begin{pmatrix}1&-1\\0&-1\end{pmatrix}$	$\begin{pmatrix}1&-1\\0&-1\end{pmatrix}$	$\begin{pmatrix}-1&-1\\0&-1\end{pmatrix}$	$\begin{pmatrix}1&1\\0&-1\end{pmatrix}$	$\begin{pmatrix}1&-1\\0&1\end{pmatrix}$	$\begin{pmatrix}-1&1\\0&-1\end{pmatrix}$	$\rightarrow$	$\begin{pmatrix}-1&-1\\0&1\end{pmatrix}$	$\begin{pmatrix}-1&1\\0&1\end{pmatrix}$
$\begin{pmatrix}-1&1\\0&-1\end{pmatrix}$	$\begin{pmatrix}-1&1\\0&-1\end{pmatrix}$	$\begin{pmatrix}1&1\\0&-1\end{pmatrix}$	$\begin{pmatrix}-1&-1\\0&-1\end{pmatrix}$	$\begin{pmatrix}-1&1\\0&1\end{pmatrix}$	$\begin{pmatrix}1&-1\\0&-1\end{pmatrix}$	$\begin{pmatrix}-1&-1\\0&1\end{pmatrix}$	$\rightarrow$	$\begin{pmatrix}1&-1\\0&1\end{pmatrix}$
$\begin{pmatrix}-1&-1\\0&-1\end{pmatrix}$	$\begin{pmatrix}-1&-1\\0&-1\end{pmatrix}$	$\begin{pmatrix}1&-1\\0&-1\end{pmatrix}$	$\begin{pmatrix}-1&1\\0&-1\end{pmatrix}$	$\begin{pmatrix}-1&-1\\0&1\end{pmatrix}$	$\begin{pmatrix}1&1\\0&-1\end{pmatrix}$	$\begin{pmatrix}-1&1\\0&1\end{pmatrix}$	$\begin{pmatrix}1&-1\\0&1\end{pmatrix}$	$\rightarrow$

Table 3.4 Transmultiplication table for implication series

According to the multiplication tables the interaction of L^- - and L^+ -operators may result in the appearance of some other L^- -operators or L^+ -operators. In this particular setting we have generated by multiplication the complete set of operators of the equivalence series. Of the 64 interactions in the multiplication table, half are forbidden. Quite naturally, no forbidden interactions occur in the corresponding transmultiplication table.
In comparison with positive logic, with its 16 operators and multiplication table with 256 squares, negative logic comprises 66 operators, with a multiplication table now consisting of 4356 squares. The complete set of matrix logic, including its positive and negative subsets, contains the empty operator $\hat{0}$ and 80 positive and negative operators; thus the complete multiplication table of matrix logic contains 6561 interactions.

Another important question to consider concerns the interaction of negative logic operators among themselves. As in positive logic, interaction of L^- -operators also contains a subset of forbidden interactions, which is much greater than in positive logic: 1050 interactions are forbidden. The total number of forbidden interactions in the set $\Omega^+ \cap \Omega^-$ is 1827, which, in addition to the 1050 forbidden interactions of negative logic, includes 777 forbidden interactions due to the interaction between the operators of positive logic and positive and negative logics. Though negative logic contributes significantly to the subset of forbidden interactions the complete logic set is nevertheless incomparably richer in possibilities than the pure positive matrix set.

Interaction of negative logic operators among themselves not only generates new negative operators, but is capable of generating positive logic. The possibility of deriving positive logic from the negative operator set makes negative logic in some sense a generalization of positive logic, and offers a way of formally incorporating the latter in the more sophisticated and more complex domain of negative logic values. This shows itself most clearly in the significant increase of affirmation and negating power in the complete logic set $\Omega^+ \cap \Omega^-$.

Subnegators

No product of operators in the set Ω^+ is capable of generating the negation operator. Because in positive logic negation is a total and non-decomposable operation, we are inclined to think that such is generally the case. However, a simple examination of general logical evaluations shows that alongside all-out complete denials intelligent processes may involve more subtle and incomplete forms of negation. Attempts to deal with this wider concept of negation are made in multivalued logic. However, these attempts are commonly postulatory in nature and are not derived from rigorously defined principles. The negative extension of logic does offer such a rigorous principle due to the possibility of deriving the negation operator as a product of either two negative operators or negative and positive operators. In addition to the single negation of positive logic in the complete logic set $\Omega^+ \cap \Omega^-$ there are 38 different products generating the negation operator $\rceil$:

$$\begin{pmatrix} 0 & 1 \\ -1 & 0 \end{pmatrix}\begin{pmatrix} -1 & 0 \\ 0 & 1 \end{pmatrix} \qquad \begin{pmatrix} 0 & 1 \\ -1 & 1 \end{pmatrix}\begin{pmatrix} -1 & 1 \\ 0 & 1 \end{pmatrix} \qquad \begin{pmatrix} 1 & 1 \\ 0 & -1 \end{pmatrix}\begin{pmatrix} 1 & 1 \\ -1 & 0 \end{pmatrix}$$

$$\begin{pmatrix} -1 & 0 \\ 0 & 1 \end{pmatrix}\begin{pmatrix} 0 & -1 \\ 1 & 0 \end{pmatrix} \qquad \begin{pmatrix} 0 & -1 \\ 1 & 1 \end{pmatrix}\begin{pmatrix} 1 & 1 \\ 0 & -1 \end{pmatrix} \qquad \begin{pmatrix} -1 & 1 \\ 0 & 1 \end{pmatrix}\begin{pmatrix} 1 & -1 \\ 1 & 0 \end{pmatrix}$$

$$\begin{pmatrix} 0 & -1 \\ 1 & 0 \end{pmatrix}\begin{pmatrix} 1 & 0 \\ 0 & -1 \end{pmatrix} \qquad \begin{pmatrix} 0 & 1 \\ -1 & -1 \end{pmatrix}\begin{pmatrix} -1 & -1 \\ 0 & 1 \end{pmatrix} \qquad \begin{pmatrix} 1 & -1 \\ 0 & -1 \end{pmatrix}\begin{pmatrix} -1 & 1 \\ -1 & 0 \end{pmatrix}$$

$$\begin{pmatrix} 1 & 0 \\ 0 & -1 \end{pmatrix}\begin{pmatrix} 0 & 1 \\ -1 & 0 \end{pmatrix} \qquad \begin{pmatrix} 0 & -1 \\ 1 & -1 \end{pmatrix}\begin{pmatrix} 1 & -1 \\ 0 & -1 \end{pmatrix} \qquad \begin{pmatrix} -1 & -1 \\ 0 & 1 \end{pmatrix}\begin{pmatrix} -1 & -1 \\ 1 & 0 \end{pmatrix}$$

$$\begin{pmatrix} 0 & -1 \\ -1 & 0 \end{pmatrix}\begin{pmatrix} -1 & 0 \\ 0 & -1 \end{pmatrix} \qquad \begin{pmatrix} 0 & -1 \\ -1 & 1 \end{pmatrix}\begin{pmatrix} -1 & -1 \\ 0 & -1 \end{pmatrix} \qquad \begin{pmatrix} -1 & -1 \\ 0 & -1 \end{pmatrix}\begin{pmatrix} 1 & -1 \\ -1 & 0 \end{pmatrix}$$

$$\begin{pmatrix} -1 & 0 \\ 0 & -1 \end{pmatrix}\begin{pmatrix} 0 & -1 \\ -1 & 0 \end{pmatrix} \qquad \begin{pmatrix} 0 & -1 \\ -1 & -1 \end{pmatrix}\begin{pmatrix} -1 & 1 \\ 0 & -1 \end{pmatrix} \qquad \begin{pmatrix} -1 & 1 \\ 0 & -1 \end{pmatrix}\begin{pmatrix} -1 & -1 \\ -1 & 0 \end{pmatrix}$$

$$\begin{pmatrix} 0 & 1 \\ 1 & 1 \end{pmatrix}\begin{pmatrix} 1 & -1 \\ 0 & 1 \end{pmatrix} \qquad \begin{pmatrix} 1 & 0 \\ -1 & 1 \end{pmatrix}\begin{pmatrix} 0 & 1 \\ 1 & 1 \end{pmatrix}$$

$$\begin{pmatrix} 1 & 1 \\ 0 & 1 \end{pmatrix}\begin{pmatrix} -1 & 1 \\ 1 & 0 \end{pmatrix} \qquad \begin{pmatrix} -1 & 1 \\ 1 & 0 \end{pmatrix}\begin{pmatrix} 1 & 0 \\ 1 & 1 \end{pmatrix}$$

$$\begin{pmatrix} 1 & 0 \\ 1 & 1 \end{pmatrix}\begin{pmatrix} 0 & 1 \\ 1 & -1 \end{pmatrix} \qquad \begin{pmatrix} 0 & 1 \\ 1 & -1 \end{pmatrix}\begin{pmatrix} 1 & 1 \\ 0 & 1 \end{pmatrix}$$

$$\begin{pmatrix} 1 & 1 \\ 1 & 0 \end{pmatrix}\begin{pmatrix} 1 & 0 \\ -1 & 1 \end{pmatrix} \qquad \begin{pmatrix} 1 & -1 \\ 0 & 1 \end{pmatrix}\begin{pmatrix} 1 & 1 \\ 1 & 0 \end{pmatrix}$$

$$\begin{pmatrix}1&-1\\1&0\end{pmatrix}\begin{pmatrix}1&0\\1&-1\end{pmatrix} \qquad \begin{pmatrix}-1&0\\1&1\end{pmatrix}\begin{pmatrix}0&-1\\1&1\end{pmatrix}$$

$$\begin{pmatrix}1&1\\-1&0\end{pmatrix}\begin{pmatrix}-1&0\\1&1\end{pmatrix} \qquad \begin{pmatrix}1&0\\1&-1\end{pmatrix}\begin{pmatrix}0&1\\-1&1\end{pmatrix}$$

$$\begin{pmatrix}-1&-1\\1&0\end{pmatrix}\begin{pmatrix}1&0\\-1&-1\end{pmatrix} \qquad \begin{pmatrix}-1&0\\-1&1\end{pmatrix}\begin{pmatrix}0&-1\\1&-1\end{pmatrix}$$

$$\begin{pmatrix}-1&1\\-1&0\end{pmatrix}\begin{pmatrix}-1&0\\-1&1\end{pmatrix} \qquad \begin{pmatrix}1&0\\-1&-1\end{pmatrix}\begin{pmatrix}0&1\\-1&-1\end{pmatrix}$$

$$\begin{pmatrix}1&-1\\-1&0\end{pmatrix}\begin{pmatrix}-1&0\\-1&-1\end{pmatrix} \qquad \begin{pmatrix}-1&0\\-1&-1\end{pmatrix}\begin{pmatrix}0&-1\\-1&1\end{pmatrix}$$

$$\begin{pmatrix}-1&-1\\-1&0\end{pmatrix}\begin{pmatrix}-1&0\\1&-1\end{pmatrix} \qquad \begin{pmatrix}-1&0\\1&-1\end{pmatrix}\begin{pmatrix}0&-1\\-1&-1\end{pmatrix}$$

The 38 products generating the negation operator $\urcorner$ employ 38 different operators, among which there are 34 operators of negative logic and 4 operators of positive logic.

Consecutive application of appropriate pairs of these operators achieves complete negation, whereas each operator alone can be considered to be a subnegator performing certain forms of partial negation. In this way we enlarge the negating power of the complete logic set, in comparison with the single negator of positive logic.

A tangible result that follows from this examination relates to the fact that in clear contrast to positive-valued logic the operator $\urcorner$ can now be generated in multiplication interactions. Consequently, its presence in the basis sets of logic ceases to be necessary. For instance, instead of the minor basis sets we can use any of the following sets, which include one of the matrix inverses:

$$\{\leftarrow^{-1} \vee\} \qquad \{\vee \rightarrow^{-1}\}$$

$$\{\vee^{-1} \leftarrow\} \qquad \{\rightarrow \vee^{-1}\}$$

$$\{\downarrow^{-1} \rightarrow\} \qquad \{\leftarrow \downarrow^{-1}\}$$

$$\{\rightarrow^{-1} \downarrow\} \qquad \{\downarrow \leftarrow^{-1}\}$$

Affirmation

There are just two ways to achieve affirmation in positive logic, by employing either the identity operator or the square of the negation operator. With the extension of logic into the negative domain the affirmation power of logic is significantly enlarged: as many as 27 different ways of deriving the identity operator can now be established. The increase in the affirmation power of logic is due to the fact that in negative logic the set of non-singular operators and therefore the set of possible inverses is further enlarged. Altogether in the complete logic set integer-valued inverses, whose elements are defined in the ternary set $E_3 = \{-1, 0, 1\}$, can be computed for 40 logic operators.

Because the complete logic set $\Omega^+ \cap \Omega^-$ contains only 33 singular operators, this leaves us with a still unaccounted-for octet of non-singular operators from the tautology series, the inverses of which cannot be expressed in the integer set of truth-values $E_3 = \{-1, 0, 1\}$. The inverses for the fundamental octet are given on the next page. The fact that logic space does not hold within itself the complete set of inverses not only illuminates the need for the further extension of logic, but also shows that logic is fundamentally incomplete without its physical component!

At this stage we are not yet ready to discuss the meaning of the fundamental octet of inverses of the tautology series. In a later section, devoted to logic quantum numbers, we shall show that these play the fundamental role of diagonalizing operators.

The usage of logic inverses has decidedly non-trivial consequences for the theory of high-level intelligence. The inverse transformation occupies a prominent place in the operations of the mind. Often the issue arises whether to use rules for forward or backward inference, in other words, whether deduction should be driven by putting questions or by the available facts. For a subset of logic with division, where the identity operation is everywhere defined, the execution of each operation induces a transformation, for which there always exists another operation such that if the two are performed one after the other we eventually return to the initial state, because every pair of adjacent operations which are inverses of each other cancels the transformation.

The general conclusion that stems from this investigation can be summarized as follows. The possibility of multiple forms of affirmation dispels the common misconception that the idea of logical equivalence is applicable only to those composition rules which involve identity or the square of the negation operator. The fact that there are many different options for affirmation, depending in part on the logic operation to be performed, is an important factor allowing an enlargement of intelligent capabilities.

L^-	$\det L^-$	$(L^-)^{-1}$
$\begin{pmatrix} 1 & 1 \\ 1 & -1 \end{pmatrix}$	-2	$\begin{pmatrix} \frac{1}{2} & \frac{1}{2} \\ \frac{1}{2} & -\frac{1}{2} \end{pmatrix}$
$\begin{pmatrix} 1 & 1 \\ -1 & 1 \end{pmatrix}$	2	$\begin{pmatrix} \frac{1}{2} & -\frac{1}{2} \\ \frac{1}{2} & \frac{1}{2} \end{pmatrix}$
$\begin{pmatrix} 1 & -1 \\ 1 & 1 \end{pmatrix}$	2	$\begin{pmatrix} \frac{1}{2} & \frac{1}{2} \\ -\frac{1}{2} & \frac{1}{2} \end{pmatrix}$
$\begin{pmatrix} -1 & 1 \\ 1 & 1 \end{pmatrix}$	-2	$\begin{pmatrix} -\frac{1}{2} & \frac{1}{2} \\ \frac{1}{2} & \frac{1}{2} \end{pmatrix}$
$\begin{pmatrix} -1 & -1 \\ -1 & 1 \end{pmatrix}$	-2	$\begin{pmatrix} -\frac{1}{2} & -\frac{1}{2} \\ -\frac{1}{2} & \frac{1}{2} \end{pmatrix}$
$\begin{pmatrix} -1 & -1 \\ 1 & -1 \end{pmatrix}$	2	$\begin{pmatrix} -\frac{1}{2} & \frac{1}{2} \\ -\frac{1}{2} & -\frac{1}{2} \end{pmatrix}$
$\begin{pmatrix} -1 & 1 \\ -1 & -1 \end{pmatrix}$	2	$\begin{pmatrix} -\frac{1}{2} & -\frac{1}{2} \\ \frac{1}{2} & -\frac{1}{2} \end{pmatrix}$
$\begin{pmatrix} 1 & -1 \\ -1 & -1 \end{pmatrix}$	-2	$\begin{pmatrix} \frac{1}{2} & -\frac{1}{2} \\ -\frac{1}{2} & -\frac{1}{2} \end{pmatrix}$

3.1.2 Vacuum interactions

We have arrived at the need to consider negative logic as a result of a generalization of the notion of a logic inverse, which became possible due to the matrix formulation of logic. This is the most logical and consistent approach, which is not initiated by chance but is forced upon us by the nature of the inverse operation in matrix algebra. However, there is a more fundamental way of developing the negative extension of logic. Within the framework of physics negative logic can be consistently interpreted as an antilogic in which true and false are recombined with each other.

The relation between logic and antilogic resembles in many respects the relationship between matter with positive levels of energy and antimatter with vacuum negative levels of energy.

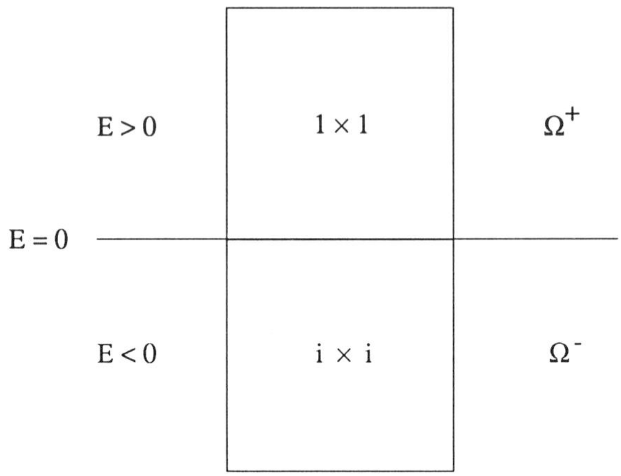

Similarly, as the Dirac theory of matter (the electron) has led to the mathematical discovery of antimatter (the positron), the operator theory of logic leads to the establishment of antilogic with its concept of vacuum logic values. If we identify a vacuum with the empty operator $\hat{0}$, then the total polarization of the vacuum, in the balanced ternary value set

$$E_3 = \{-1, 0, 1\} ,$$

will form precisely the 80 logic operators, whose sum is the empty operator,

$$\hat{0} = \sum_{i}^{80} L_i \, .$$

A natural condition for the conservation of the total logic charge is satisfied.

It is necessary to emphasize that negative logic should not be perceived as a simple mirror inversion of positive logic. The relationship between the two logics is more complex. Contrary to what one might expect negative logic is not the complement of discrete positive logic but the complement of continuous logic. The negative truth-value –1 is derived by applying the complement operator to the extended alphabet of truth-values

$$\{0, 1, 2\} ,$$

which we arrived at in continuous matrix logic:

x	$\overline{x}$
0	1
1	0
2	–1
–1	2

Consequently, matrix operator logic establishes altogether 4 fundamental truth-values:

$$\{-1, 0, 1, 2\}\ .$$

Positive logic consists mostly of non-commuting operators. Though this fact is important on its own merits, our concern is here not with the commutation of logic operators but with the fact that products of non-commutation, with the exception of inhibition and converse inhibition, are negative logic operators. Whereas Ω^+ -logic is generated from Ω^- -logic by means of multiplication, Ω^- -logic can be generated from Ω^+ -logic by subtraction. Any negative operator L_k^- can be expressed via a subtraction of two positive operators

$$L_k^- = L_i^+ - L_j^+$$

with the corresponding bracket product taking the form:

$$<x \mid L_k^- \mid y> = <x \mid L_i^+ - L_j^+ \mid y> = <x \mid L_i^+ \mid y> - <x \mid L_j^+ \mid y>\ .$$

The problem of negative logic is quantized into two problems of two positive logics bonded to each other by subtraction of the appropriate energy levels.
Since

$$L_i^+ - L_j^+ = \overline{L_j^+} - \overline{L_i^+}$$

the bracket product can also be expressed as

$$<x \mid L_k^- \mid y> = \overline{<x \mid L_j \mid y>} - \overline{<x \mid L_i \mid y>}\ .$$

Comparison operators

Among the 81 operators of complete logic there are only two antisymmetric operators

$$\tilde{L} = -L ,$$

which fulfil the fundamental function of comparison:

$$\begin{pmatrix} 0 & -1 \\ 1 & 0 \end{pmatrix} \quad \text{and} \quad \begin{pmatrix} 0 & 1 \\ -1 & 0 \end{pmatrix}.$$

Within the negative extension of logic the operators of comparison have clear logical meaning. We obtain for the first operator

$$< p \begin{pmatrix} 0 & -1 \\ 1 & 0 \end{pmatrix} q > \; = p - q$$

and for the second

$$< p \begin{pmatrix} 0 & 1 \\ -1 & 0 \end{pmatrix} q > \; = q - p .$$

The operators compute the difference between the truth values. In case of binary truth values the results of comparison fall into the ternary set

$$E_3 = \{-1, 0, 1\}$$

Clearly, the antisymmetric comparison operators are matrix transposes of each other:

$$\widetilde{\begin{pmatrix} 0 & -1 \\ 1 & 0 \end{pmatrix}} = \begin{pmatrix} 0 & 1 \\ -1 & 0 \end{pmatrix},$$

and therefore

$$< p \begin{pmatrix} 0 & -1 \\ 1 & 0 \end{pmatrix} q > \; = - < p \begin{pmatrix} 0 & 1 \\ -1 & 0 \end{pmatrix} q > .$$

In view of the importance of the comparison operators we shall reserve for them special symbols:

$$\blacktriangledown = \begin{pmatrix} 0 & -1 \\ 1 & 0 \end{pmatrix}$$

$$\blacktriangle = \begin{pmatrix} 0 & 1 \\ -1 & 0 \end{pmatrix},$$

where ▼ is the comparison operator and ▲ the converse comparison operator. Comparison and converse comparison are orthogonal operators

$$\tilde{\blacktriangledown} = \blacktriangledown^{-1}$$

$$\tilde{\blacktriangle} = \blacktriangle^{-1},$$

and inverses of each other

$$\blacktriangledown\,\blacktriangle = \blacktriangle\,\blacktriangledown = I\,.$$

The fundamental operators of comparison indicate a clear connection between the operators of negative logic and those of continuous logic. The comparison operators can be derived by complementation from two operators available in a continuous universe:

$$\overline{<p\,|\rightarrow^2|\,q>} = <p\,\overline{\begin{pmatrix} 1 & 2 \\ 0 & 1 \end{pmatrix}}\,q> = <p\begin{pmatrix} 0 & -1 \\ 1 & 0 \end{pmatrix}q> = <p\,|\blacktriangledown|\,q>$$

$$\overline{<p\,|\leftarrow^2|\,q>} = <p\,\overline{\begin{pmatrix} 1 & 0 \\ 2 & 1 \end{pmatrix}}\,q> = <p\begin{pmatrix} 0 & 1 \\ -1 & 0 \end{pmatrix}q> = <p\,|\blacktriangle|\,q>$$

The fundamental role of comparison operations for negative logic is illuminated by the fact that the antisymmetric comparison operators allow us to define, in addition to genuinely scalar truth values, a special subclass of logic values which are pseudoscalars. With a change-over from the left coordinate system to the right coordinate system a pseudoscalar remains invariant in absolute value, but changes its sign to the opposite:

$$<x \mid \blacktriangledown \mid y> \;=\; -<y \mid \blacktriangledown \mid x> \;=\; -<x \mid \blacktriangle \mid y> .$$

The operation of comparison is of great importance in logic evaluations. The extension of logic into the negative domain enables us to express the operation of comparison rigorously in terms of the antisymmetric matrix logic operators.

3.1.3 Chronologic

The meaning of negative logic is greatly clarified through the links between negative truth-values and time. Time is an important aspect of logic evaluation. Ignoring the time factor may destroy causal relations and invalidate logic operations that are otherwise perfectly correct. Though time plays a fundamental role in the logic processes of deduction and induction, the question remains unsolved of how to include the time factor explicitly in logic equations. In this section we investigate the coupling in time of logic states, which opens new avenues for chronologic and clarifies the meaning of negative truth-values.

Consider a logic state at the instant of time τ_1, which tunnels into a new state of a higher truth status:

$$| q (\tau_1)> \;\Rightarrow\; | q (\tau_2)>$$

where

$$| q (\tau_1) > = \begin{pmatrix} \bar{q} \\ q \end{pmatrix} \qquad \text{and} \qquad | q (\tau_2) > = \begin{pmatrix} \bar{q} - \Delta \\ q + \Delta \end{pmatrix}$$

Such transitions can characterize the creative operations of the mind, when the verum component of the logic state vector increases by a certain value Δ, while due to the logic normalization condition the falsum component simultaneously decreases by the same value Δ.

If the instant of time τ_2 precedes the instant of time τ_1, tunnelling $+\Delta$ of a logic state forwards in time is equivalent to tunnelling $-\Delta$ backwards in time. In a time map this can be depicted as follows:

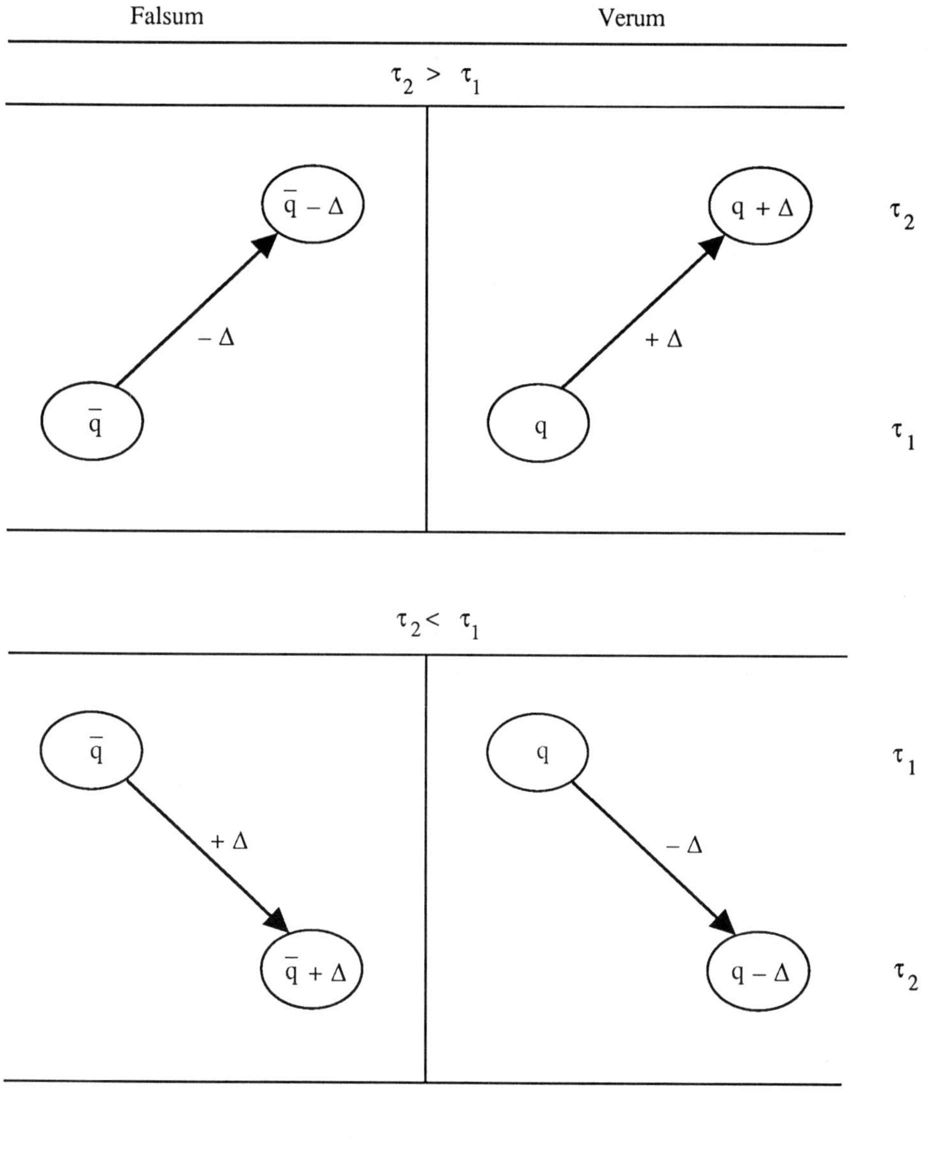

When time runs forward, $\tau_2 > \tau_1$, the verum component increases from q to q + Δ. When the direction of time is reversed, $\tau_1 > \tau_2$, the verum component decreases from q to q – Δ:

$$\tau_2 > \tau_1: \quad (\bar{q}, q) \Rightarrow (\bar{q} - \Delta, q + \Delta)$$

$$\tau_2 < \tau_1: \quad (\bar{q}, q) \Rightarrow (\bar{q} + \Delta, q - \Delta)$$

In view of the normalization condition, transformation of the falsum components is reciprocal to that of the verum components.

When we consider logic states as depending explicitly on time, tunnelling backwards in time, which decreases the verum component by $-\Delta$, is equivalent to tunnelling $+\Delta$ forwards in time. Consequently, what we have for verum is reversed for falsum: positive tunnelling forwards in time for verum implies negative tunnelling backwards in time for falsum and vice versa.

To put our consideration on a more mathematical footing, we employ fundamental comparison operators, which make it possible to obtain a measure for Δ. Because logic states become explicit functions of time, their position in the bracket product is essential. Normally in matrix logic the antecedent occupies the left position in the bracket product, and the consequent the right position. For the convenience of a reader with a physical background we shall employ reversed positions for the antecedent and the consequent. This has no bearing on the results, which can be formulated equally well in the standard order of matrix logic.

Forward direction of time

$$\boxed{\tau_2 > \tau_1}$$

The instant of time τ_2 is later than the instant of time τ_1. The verum component q at instant τ_1 becomes $q + \Delta$ at instant τ_2. Then

$$\overset{\tau_2}{<q+\Delta} \,|\, \blacktriangledown \,|\, \overset{\tau_1}{q>} \;=\; q+\Delta-q \;=\; \Delta$$

This equation provides a measure of the tunnelling of the verum component forwards in time.

Backward direction of time

$$\boxed{\tau_1 > \tau_2}$$

The instant of time τ_2 is earlier than the instant of time τ_1.

The time reversal is achieved by a transposition of the comparison operator

$$\tilde{\blacktriangledown} = \blacktriangle ,$$

which produces the converse comparison operator. Then we obtain

$$\overset{\tau_1}{<}q+\Delta\,|\,\blacktriangle\,|\,\overset{\tau_2}{q}> \;=\; q-(q+\Delta) \;=\; -\Delta$$

This equation describes the tunnelling of the verum component backwards in time.
With tunnelling forwards in time Δ verum is created, with tunnelling backwards in time Δ verum is annihilated. The processes of tunnelling forwards and backwards in time are fully symmetric and differ only in their sign. But precisely the same symmetry exists between the verum and falsum components in a given direction of time.
Consequently, we can obtain the transformation of the falsum component in the forward direction of time from the transformation of the verum component in the backward direction of time and vice versa. For that purpose we make use of the combined operation of symmetry θ^2, which on the one hand permits us to pass from the basis v to the complementary basis $\overline{v}$ and on the other hand to induce the orthogonal transformation of the comparison operator.
We achieve our objective because a transition to the complementary basis recombines the verum and falsum components, while the orthogonal transformation of comparison operators induces their transposition, which implies the inversion of the direction of time.

<u>Falsum</u> <u>Forward direction of time</u>

$$\tau_2 > \tau_1$$

$$\begin{aligned}\overset{\tau_1}{<}q+\Delta\,|\,\blacktriangle\,|\,\overset{\tau_2}{q}> &= <q+\Delta\,|\,\theta^2\,\blacktriangle\,|\,q> = \overset{\tau_2}{<}\overline{q+\Delta}\,|\,\blacktriangledown\,|\,\overset{\tau_1}{\overline{q}}> = \\ &= <\overline{q}-\Delta\,|\,\blacktriangledown\,|\,\overline{q}> = \\ &= \overline{q}-\Delta-\overline{q} = -\Delta\end{aligned}$$

■

<u>Falsum</u> <u>Backward direction of time</u>

$$\tau_1 > \tau_2$$

$$\begin{aligned}\overset{\tau_2}{<}q+\Delta\,|\,\blacktriangledown\,|\,\overset{\tau_1}{q}> &= <q+\Delta\,|\,\theta^2\,\blacktriangledown\,|\,q> = \overset{\tau_1}{<}\overline{q+\Delta}\,|\,\blacktriangle\,|\,\overset{\tau_2}{\overline{q}}> = \\ &= <\overline{q}-\Delta\,|\,\blacktriangle\,|\,\overline{q}> = \\ &= \overline{q}-(\overline{q}-\Delta) = \Delta\end{aligned}$$

■

In chronologic, inversion of the direction of time recombines the verum and falsum components, just as the negation operation does. Conversely, the complementation or negation of a logic state gives an inverse reading as to the direction of time. Consequently, a physical transformation may condition the outcome of logic operations, while the logic operation may condition the direction of a time transformation.

These considerations provide important clues for the understanding of high-level intelligence. Chronologic offers a plausible analysis of causal and acausal properties of the mind, which has the fundamental capability of forward and backward inference, just as the relativistic fields split into two, positive and negative, branches, for particle and antiparticle respectively.

In addition to the logic coordinates verum and falsum, the logic states must also be characterized by an appropriate sign associated with the causal time coordinate. In this way chronologic proves to be useful in accounting for the creative unpredictability of the mind, and indicates the quantum-relativistic roots of intelligence phenomena.

The intelligent processes comprise causal and acausal imaginary logics, which are divided by the relativistic spacetime cone and can interfere with each other only within the limits imposed by the uncertainty principle:

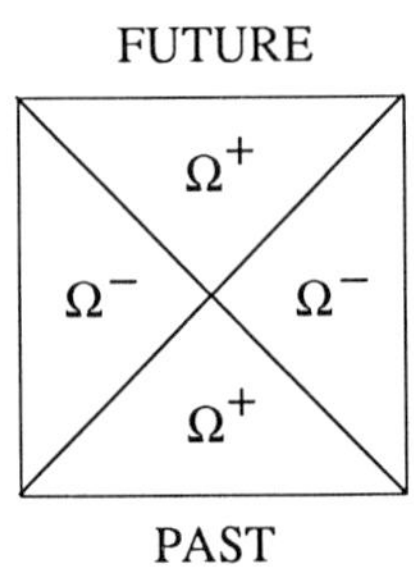

Acausal antilogic is linked with the spacelike region of the cone, and causal positive logic with the timelike region.

3.2 Artificial intelligence

The goal of bringing logic fully into the domain of the exact sciences can be achieved only if logic theory can account adequately for real-world phenomena, both natural and artificial. The field of artificial intelligence provides a perfect testing ground for a new logic theory. The role of logic in the field of artificial intelligence can be compared to the role of analysis in physics and engineering.

Artificial intelligence seeks to expand the boundaries of computer science in order to solve problems traditionally considered to require human ingenuity. The ultimate goal is to design computing systems that will reason and behave more like an intelligent mind than an ordinary number-cruncher. However, progress in the development of artificial intellect is greatly impaired by the lack of clear understanding of the gap that exists between the logic of the inductive mind and those logics employed in a computer.

In current practice, the basic functions of a computer are accomplished by devices that have two stable states, one of the states representing 1, the other state representing 0, thus allowing implementation of the binary number system. The computer problems of artificial intelligence can be expressed as a sequence of operations on logic functions, making use of the basic operation addition mod 2. Any artificial intelligence problem, either simple or complex, can in the end be reduced to a sequence of binary operations in the mod 2 number system.

Contemporary theories of artificial intelligence seek primarily to develop elaborate programs which are supposed to duplicate high-level intelligence. The numerous failures of this approach have given rise to the idea that the puzzles of the mind are more likely to be unlocked through a deeper exploration of logic. With the development of matrix operator logic, the search for the rules of the mind has received a new impetus.

Matrix logic extends the power of direct mathematical computation to the complete set of logic functions. At the same time it establishes a special subset of inadmissible products for logic operators, termed the forbidden interactions. The presence of inadmissible products is a distinct feature of real-world logic, which fundamentally distinguishes it from processes occurring in the artificial intelligence systems realized by computers.

3.2.1 Forbidden interactions

The proper meaning of intelligent operations is often defined not by what the mind can do, but by what is precluded from doing. Due to physical constraints, the scope of logic operations is restricted by the set of forbidden interactions. This fundamental aspect of natural intelligence is not accounted for in artificial systems because of the limitations imposed by operations in the mod 2 number system. Clearly, the introduction of the mod 2 number system prevents the appearance of the forbidden logic values. And the multiplication table of matrix logic thus becomes as follows:

×	↔	⌉	∧	∨	→	←	↓	↑	\	/	⊥	T	⊣	⊢	1	0
↔	↔	⌉	∧	∨	→	←	↓	↑	\	/	⊥	T	⊣	⊢	1	0
⌉	⌉	↔	/	→	∨	↓	←	\	↑	∧	T	⊥	⊣	⊢	1	0
∧	∧	\	∧	⊥	∧	⊥	\	0	\	0	⊥	0	∧	\	⊥	0
∨	∨	←	⊣	↓	⌉	→	↔	\	⊢	∧	1	⊥	/	↑	T	0
→	→	↓	⊣	←	↔	∨	⌉	↑	⊢	/	1	T	∧	\	⊥	0
←	←	∨	∧	⌉	↓	↔	→	⊢	\	⊣	⊥	1	/	↑	T	0
↓	↓	→	/	↔	←	⌉	∨	⊢	↑	⊣	T	1	∧	\	⊥	0
↑	↑	/	0	/	T	↑	T	↑	0	/	0	T	/	↑	T	0
\	\	∧	0	∧	⊥	\	⊥	\	0	∧	0	⊥	∧	\	⊥	0
/	/	↑	/	T	/	T	↑	0	↑	0	T	0	/	↑	T	0
⊥	⊥	⊥	∧	\	\	∧	∧	\	\	∧	⊥	⊥	0	0	0	0
T	T	T	/	↑	↑	/	/	↑	↑	/	T	T	0	0	0	0
⊣	⊣	⊢	⊣	1	⊣	1	⊢	0	⊢	0	1	0	⊣	⊢	1	0
⊢	⊢	⊣	0	⊣	1	⊢	1	⊢	0	⊣	0	1	⊣	⊢	1	0
1	1	1	⊣	⊢	⊢	⊣	⊣	⊢	⊢	⊣	1	1	0	0	0	0
0	0	0	0	0	0	0	0	0	0	0	0	0	0	0	0	0

Table 3.5 Addition mod 2, 1 + 1 = 0, multiplication table

The mod 2 multiplication table coincides with the standard table, except for the subset of forbidden interactions, which are now completely eliminated. In their place, it now becomes possible to generate six logic operators {↔, ⌉, ∨, →, ←, ↓}. From the viewpoint of artificial intelligence this could be interpreted as an advantage. In fact, mod 2 logic definitely amputates those capabilities of logic which are essential for inductive inference and genuine creativity. The total elimination of forbidden interactions is not only in conflict with the physical foundations of the intelligence code, it also totally annuls the continuous and negative components of logic!

A clear advantage of mod 2 logic from the computational point of view is that its operations could never transcend the initial operator set. However, this advantage, alongside the increase in the number of possible interactions that is obtained through the cancellation of forbidden products, is achieved at the high cost of eliminating the powerful set of operators from continuous and negative logics.

Consider the compound operator (∨ + ∧) in a continuous universe. In the mod 2 system the operator is replaced by the negation operator:

$$\text{mod } 2: \qquad \vee + \wedge = \begin{pmatrix} 0 & 1 \\ 1 & 1+1 \end{pmatrix} = \begin{pmatrix} 0 & 1 \\ 1 & 0 \end{pmatrix} = \urcorner .$$

On the other hand, in standard logic we had:

$$\text{mod } 10: \qquad \vee - \wedge = \urcorner .$$

This shows the equivalence of positive and negative unit values in the mod 2 system, where –1 is equivalent to +1:

$$\text{mod } 2: \qquad \begin{aligned} 1 + 1 &= 0 , \\ 1 - 1 &= 0 . \end{aligned}$$

Because in mod 2 matrix logic –1 operates exactly as +1, we can simply replace negative values by positive ones. This consequently cancels the negative logic, thereby destroying a crucial aspect of intelligent capabilities. The elimination of negative values implies the elimination of chronologic. Thus, an important component of intelligence is amputated in artificial intelligence systems.

The set of 81 operators of the complete logic $\Omega^+ \cap \Omega^-$ is reduced to the initial basic set of just 16 logic operators. Instead of a rich set of 48 non-singular logic operators we are back to square one with just the 6 non-singular operators of the initial basic set:

$$\{\leftrightarrow, \urcorner, \vee, \rightarrow, \leftarrow, \downarrow\} .$$

In view of the equivalence of positive and negative values in a mod 2 system, implication and converse implication now become inverses of each other:

$$\text{mod } 2: \qquad \begin{aligned} \rightarrow\rightarrow &= I \\ \leftarrow\leftarrow &= I \end{aligned}$$

This is in striking contrast with the products in standard logic:

$$\text{mod } 10: \qquad \begin{pmatrix} 1 & -1 \\ 0 & 1 \end{pmatrix}\begin{pmatrix} 1 & 1 \\ 0 & 1 \end{pmatrix} = I$$

$$\begin{pmatrix} 1 & 0 \\ -1 & 1 \end{pmatrix}\begin{pmatrix} 1 & 0 \\ 1 & 1 \end{pmatrix} = I$$

The operators of disjunction $\vee$ and NAND, which have their inverses in the negative subset of logic, now become inverses of each other:

$$\text{mod 2:} \qquad \vee^{-1} = \downarrow , \qquad \downarrow^{-1} = \vee .$$

In the complete logic $\Omega^{+} \cap \Omega^{-}$ we are, through real physical interactions, able to generate the negation operator $\rceil$ in a great many ways. This is replaced in the artificial mod 2 system by just four products:

$$\text{mod 2:} \qquad \vee \rightarrow \; = \; \downarrow \leftarrow \; = \; \leftarrow \vee \; = \; \rightarrow \downarrow \; = \; \rceil .$$

The octet of different interactions associated in standard logic with each of the above products are all cancelled in artificial systems. The multiple ways of generating the negation operator are greatly reduced.

What occurs in artificial intelligence systems can rightly be defined as a destruction of the valuable assets of the intelligence code. Admittedly, the mod 2 system makes it possible to generate the negation and equivalence operators; yet in comparison with the antilogic, where the generation of these operators is physically meaningful, in the mod 2 system it is in fundamental disagreement with the basic rules of the intelligence code.

Clearly, the computer is a gross simplification of the mind because the physical foundation of logic is thrown out. It seems that the technical availability of devices with 2 stable states, as compared with the difficulties of physically realizing devices with 4 natural states (the fundamental alphabet), has misled us excessively. Computer logic does not solve but rather fakes the properties of negative logic for artificial intelligence. The ultimate artificial intelligence systems will have to incorporate negative and continuous logics, instead of the convenient device of mod 2. The development of machine intelligence is as much an engineering problem as it is a fundamental problem. The concept of forbidden logic interactions makes it clear that further advances in artificial intelligence can be achieved only through the development of a unified theory of physics and logic.

3.2.2 Autoproducts

This section is included not only because of its intrinsic interest, but also, and more importantly, to emphasize the importance of matrix logic for modelling and understanding intelligence, both natural and artificial, computationally. In any fruitful generalization of matrix logic it is desirable to retain the useful features of the basic theory, just as matrix logic itself incorporates conventional logic as a scalar limit. One naturally expects a new logic theory to be able to account computationally for the properties of the mind which conventional logic is unable to either predict or explain. Though the simplicity of the suggestion that intelligence is just computation may insult the complexity of the mind, the dream of an intelligent machine is widely shared by computer scientists.

However, research has determined that conventional logic is inefficient and inadequate to incorporate high-level intelligence processing. Whereas conventional computing excels at numerical problems, it has not been effective in dealing with problems that require creative intelligence. It became apparent that to come to grips with these fundamental problems, a deeper exploration of logic is required.

Matrix logic calls for a re-examination of fundamental issues: it offers viable solutions for overcoming limitations in current knowledge processing technology, and deepens our understanding of the intelligence code. Matrix constructs are an important tool of the developed intellect. The scientific intellect in particular tends to organize data in matrix form whenever possible. The formation of matrix structures such as data matrices or graphs is an essential coding step in abstract thought, preceding analytical extrapolation. While an abstract thought has all the properties of a scalar, there is strong evidence that its codification is based on manipulating vector and matrix quantities. Therefore, high-level intelligence, whether natural or artificial, must be able to accept both scalar and vector codes.

This brings us to the fundamental question of the integration of processing and memory functions in natural intelligent systems. The fundamental distinction between the architectures of the mind and that of the computer concerns the organization of memory and processing. In a conventional computer the processor and memory units are separate systems linked by communication channels. In contrast, the mind can be defined as the memory processor, where the memory function and the processing function are integrated and inseparable.

In the absence of a proper mathematical language capable of unifying the memory and processing operations we cannot hope to duplicate the power of the intelligent mind. This is closely related to an issue brought into focus by matrix logic. Whereas in traditional computing theory a division into operators and operands is essential, in the matrix logic formalism any true distinction between the operator and operands disappears.

The failures of artificial intelligence must be attributed not only to relatively poor knowledge of the rules of high-level intelligence, but even more importantly to the absence of adequate logic structures capable of accommodating the fundamental integration of the memory function and the processing function. The development of matrix operator logic opens new avenues for duplicating the logic of the intelligence code in artificial systems.

This brings us to the question of the autonomous character of intelligent operations. The autonomous capability allows intelligent systems to postpone, if necessary, *instant* responses to input from the environment and to carry out necessary evaluations prior to actions. In mathematical terms the autonomous capability of the mind implies a closed topology which can be associated with closed logic structures.

The expressions in closed logic are constructed from matrix strings by connecting the end ket vector of a string with the front bra vector. The closed logic expression thus takes the form of a circular string, defined as the autoproduct:

$$\overbrace{\langle\,|\,|\,\rangle\langle\,|\,|\,\rangle\langle\,|\,|\,\rangle\ldots\langle\,|\,|\,\rangle}^{\times}$$

The concept of an autoproduct raises the possibility of integrating the memory and processing functions in the same logic structure, the structure which can serve as a blueprint for the *memory processor* of artificial intelligence systems.

Autoproducts are characterized by interesting mathematical properties. Whereas the evaluation or reading of a logic expression, in the fundamental product representation, can begin from any point and spread in either direction, to the left or to the right or both, in the autoproduct the result of computation depends on the point at which the reading of an expression begins, just as the proper decoding of genetic information requires a correct initialization point.

Being accustomed to the fact that the associativity of multiplication holds universally, one does not question whether it is indeed applicable with respect to autonomous intelligent operations. A surprising novelty brought about by closed logic consists in its creating a new kind of matrix relation in which the associativity law can no longer be maintained. Furthermore, in contrast to linear logic, where the result of computation is a scalar value, in closed logic an expression can evaluate to either a scalar, which is the same should a closed string decay into an open one, or to a set of logic operators, depending on the point from which the computation has begun. If the initialization point is a logic operator or ket logic vector, the result of computation is a logic operator, but if it is a bra logic vector, then the result is a scalar, identical for the evaluation of the corresponding fundamental product. Thus, in closed logic the associativity law surrenders to a more general principle, according to which the mathematical nature of the product, surprisingly enough, becomes dependent on the order of operations!

The autoproduct is clearly an object with unusual mathematical properties. In an attempt to answer the question what it actually represents, a scalar or an operator, we are confronted with a situation that is similar in some sense to the wave-particle dualism in quantum physics, where in one measurement condition a quantum process manifests particle properties, but in another measurement condition, wave properties. Similarly, the autoproduct is neither a scalar nor an operator, but both a scalar and an operator depending on whether the internal or external mode of reasoning is required.

The interiorization and exteriorization processes lend themselves most naturally to treatment by autoproducts and together with the non-associativity of closed logic provide important clues to the puzzle of high-level intelligence. The duality principle of closed logic described above allows an intelligent system to realize congruently two modes of operation: the scalar mode, associated with the processing function, and the operator mode, associated with the memory function.

Since ideally an intelligent machine must be able to accept both scalar and matrix signals, a formidable goal of artificial intelligence can thus be formulated as the design of a non-associative logic machine.

In contrast with conventional computing systems, which process operands, the autonomous machine must have the capacity of operating on both operands and operators. In fact, in an autonomous intelligent computer, not only scalar operands but also operators can serve as input signals, transmitted for further evaluations. Consider the equivalence gate in the processing mode and in the operator memory mode:

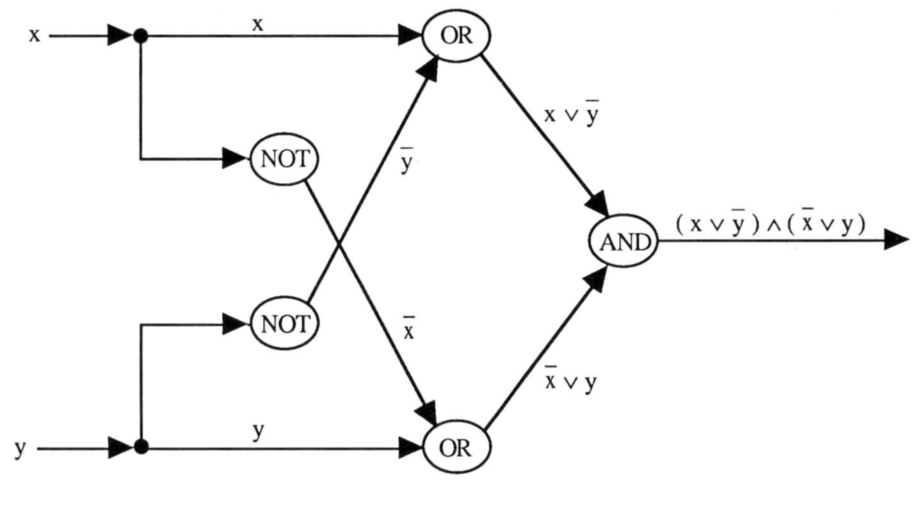

Conventional equivalence gate

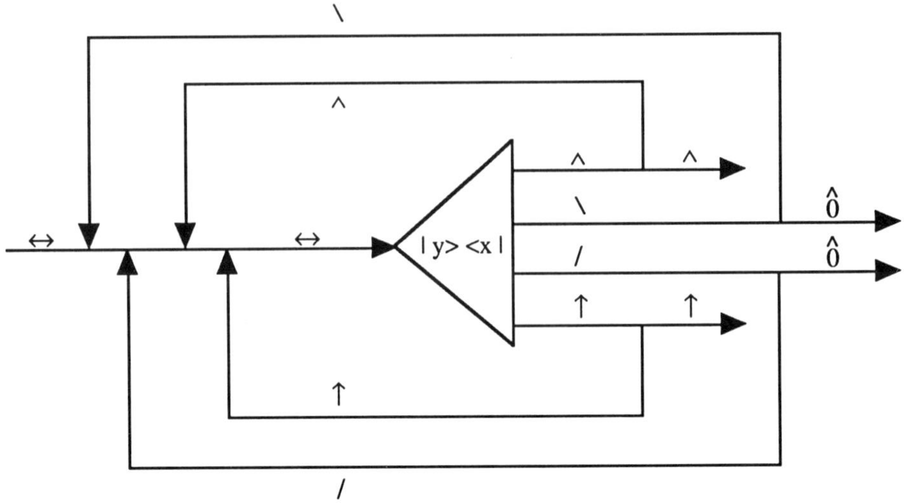

Autoproduct equivalence gate

The conventional gate evaluates the equivalence function $\langle x | \leftrightarrow | y\rangle$. It receives scalar values and outputs scalar values. Upon execution of the operation, everything is forgotten, nothing (no memory) remaining.

The autoproduct equivalence gate is a circuit with a memory, which can evaluate the equivalence function while in the scalar mode, and can compute the memory autoproduct while in the operator mode. Mathematically the operator mode can be described as follows:

$$\overset{\ulcorner \times \urcorner}{\langle x | L | y\rangle} = (L\,|y\rangle\langle x|)^{n},$$

where n is the number of iterations.

In a discrete universe there are 4 outcomes for the outer product $|y\rangle\langle x|$:

$$|y\rangle\langle x| = \begin{cases} |1\rangle\langle 1| = \wedge \\ |1\rangle\langle 0| = \backslash \\ |0\rangle\langle 1| = / \\ |0\rangle\langle 0| = \uparrow \end{cases},$$

which implies that the autoproducts for the basic logic functions may have four outputs. Below we present the autoproducts for the basic set of logic functions.

∧:

<table>
<tr><th colspan="2">Operator mode</th><th>Scalar mode</th></tr>
<tr><th>n = 1</th><th>n > 1</th><th><x | ∧ | y></th></tr>
<tr><td>∧ ∧ = ∧</td><td>∧</td><td>1</td></tr>
<tr><td>∧ \ = \</td><td>0</td><td>0</td></tr>
<tr><td>∧ / = 0</td><td>0</td><td>0</td></tr>
<tr><td>∧ ↑ = 0</td><td>0</td><td>0</td></tr>
</table>

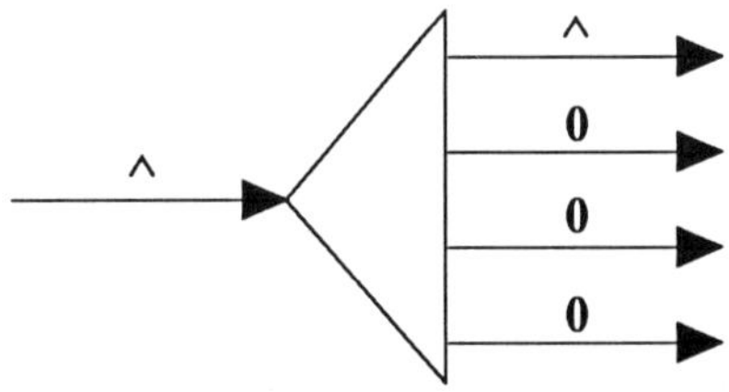

\:

<table>
<tr><th colspan="2">Operator mode</th><th>Scalar mode</th></tr>
<tr><th>n = 1</th><th>n > 1</th><th><x | \ | y></th></tr>
<tr><td>\ ∧ = 0</td><td>0</td><td>0</td></tr>
<tr><td>\ \ = 0</td><td>0</td><td>0</td></tr>
<tr><td>\ / = ∧</td><td>∧</td><td>1</td></tr>
<tr><td>\ ↑ = \</td><td>0</td><td>0</td></tr>
</table>

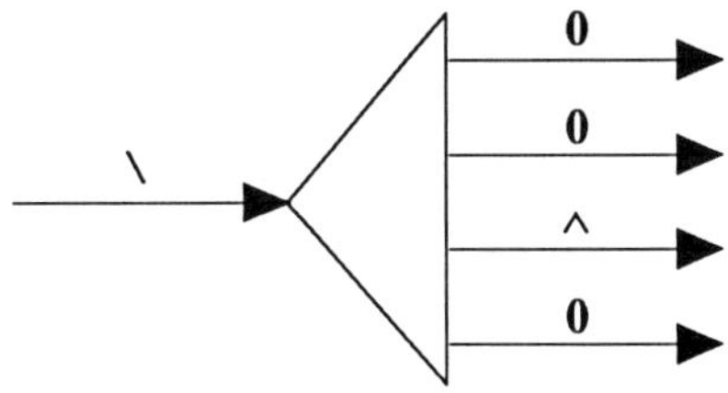

/:

<table>
<tr><th colspan="2">Operator mode</th><th>Scalar mode</th></tr>
<tr><th>n = 1</th><th>n > 1</th><th><x | / | y></th></tr>
<tr><td>/ ∧ = /</td><td>0</td><td>0</td></tr>
<tr><td>/ \ = ↑</td><td>↑</td><td>1</td></tr>
<tr><td>/ / = 0</td><td>0</td><td>0</td></tr>
<tr><td>/ ↑ = 0</td><td>0</td><td>0</td></tr>
</table>

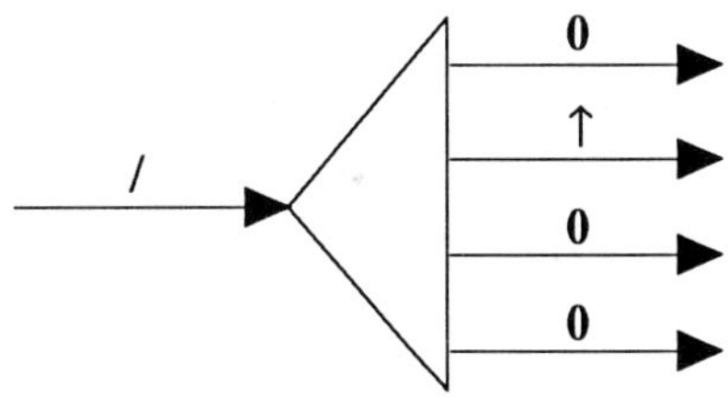

↑:

<table>
<tr><th colspan="2">Operator mode</th><th>Scalar mode</th></tr>
<tr><th>n = 1</th><th>n > 1</th><th><x | ↑ | y></th></tr>
<tr><td>↑ ∧ = 0</td><td>0</td><td>0</td></tr>
<tr><td>↑ \ = 0</td><td>0</td><td>0</td></tr>
<tr><td>↑ / = /</td><td>0</td><td>0</td></tr>
<tr><td>↑ ↑ = ↑</td><td>↑</td><td>1</td></tr>
</table>

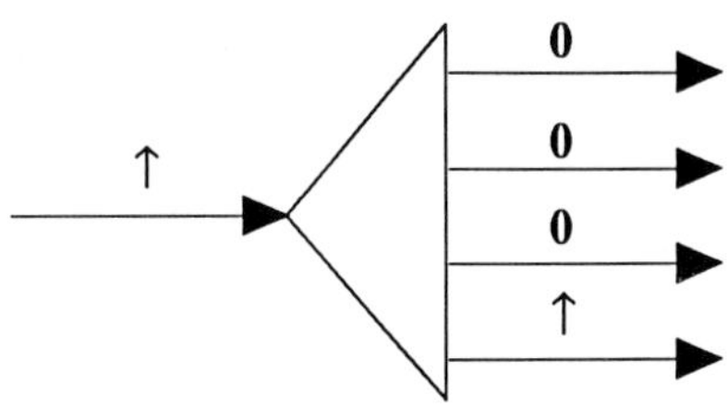

↔ :

<table>
<tr><th colspan="2">Operator mode</th><th>Scalar mode</th></tr>
<tr><th>n = 1</th><th>n > 1</th><th><x | ↔ | y></th></tr>
<tr><td>↔ ∧ = ∧</td><td>∧</td><td>1</td></tr>
<tr><td>↔ \ = \</td><td>0</td><td>0</td></tr>
<tr><td>↔ / = /</td><td>0</td><td>0</td></tr>
<tr><td>↔ ↑ = ↑</td><td>↑</td><td>1</td></tr>
</table>

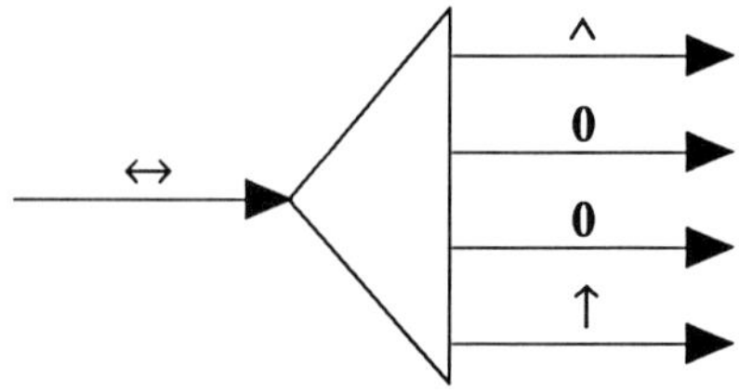

⌉ :

<table>
<tr><th colspan="2">Operator mode</th><th>Scalar mode</th></tr>
<tr><th>n = 1</th><th>n > 1</th><th><x | ⌉ | y></th></tr>
<tr><td>⌉ ∧ = /</td><td>0</td><td>0</td></tr>
<tr><td>⌉ \ = ↑</td><td>↑</td><td>1</td></tr>
<tr><td>⌉ / = ∧</td><td>∧</td><td>1</td></tr>
<tr><td>⌉ ↑ = \</td><td>0</td><td>0</td></tr>
</table>

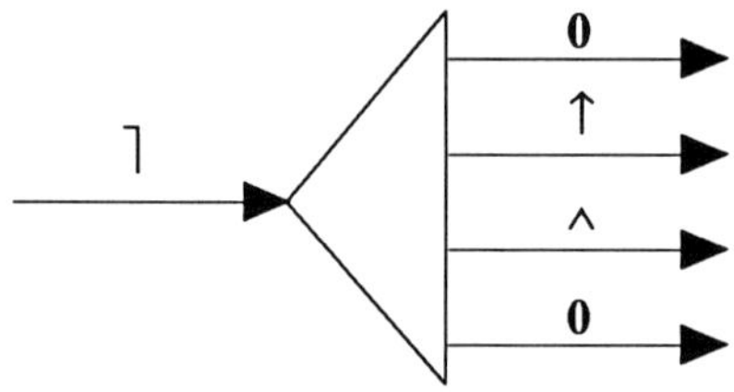

⊥ :

<table>
<tr><th colspan="2">Operator mode</th><th>Scalar mode</th></tr>
<tr><th>n = 1</th><th>n > 1</th><th><x | ⊥ | y></th></tr>
<tr><td>⊥ ∧ = ∧</td><td>∧</td><td>1</td></tr>
<tr><td>⊥ \ = \</td><td>0</td><td>0</td></tr>
<tr><td>⊥ / = ∧</td><td>∧</td><td>1</td></tr>
<tr><td>⊥ ↑ = \</td><td>0</td><td>0</td></tr>
</table>

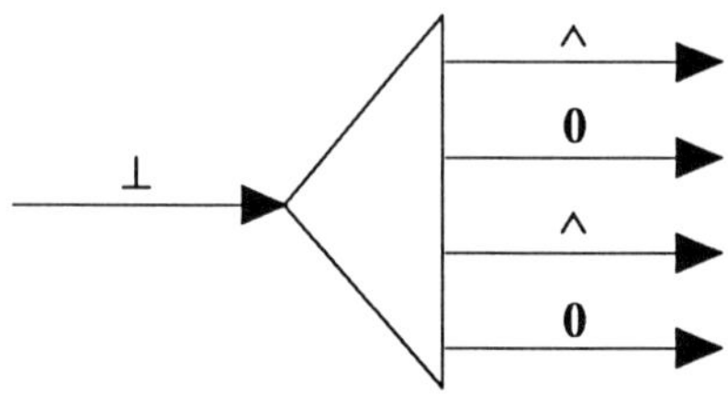

T :

<table>
<tr><th colspan="2">Operator mode</th><th>Scalar mode</th></tr>
<tr><th>n = 1</th><th>n > 1</th><th><x | T | y></th></tr>
<tr><td>T ∧ = /</td><td>0</td><td>0</td></tr>
<tr><td>T \ = ↑</td><td>↑</td><td>1</td></tr>
<tr><td>T / = /</td><td>0</td><td>0</td></tr>
<tr><td>T ↑ = ↑</td><td>↑</td><td>1</td></tr>
</table>

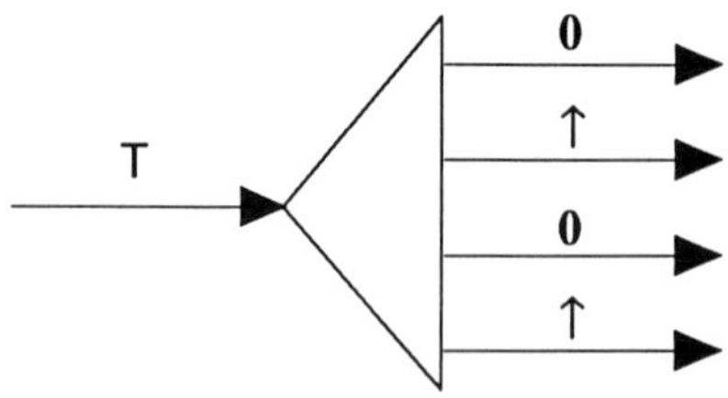

⊣ :

Operator mode		Scalar mode
n = 1	n > 1	<x | ⊣ | y>
⊣ ∧ = ⊣	⊣	1
⊣ \ = ⊢	⊢	1
⊣ / = **0**	**0**	0
⊣ ↑ = **0**	**0**	0

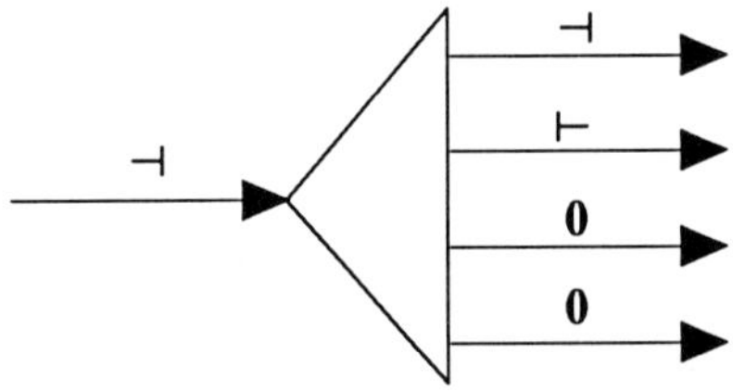

⊢ :

Operator mode		Scalar mode
n = 1	n > 1	<x | ⊢ | y>
⊢ ∧ = **0**	**0**	0
⊢ \ = **0**	**0**	0
⊢ / = ⊣	⊣	1
⊢ ↑ = ⊢	⊢	1

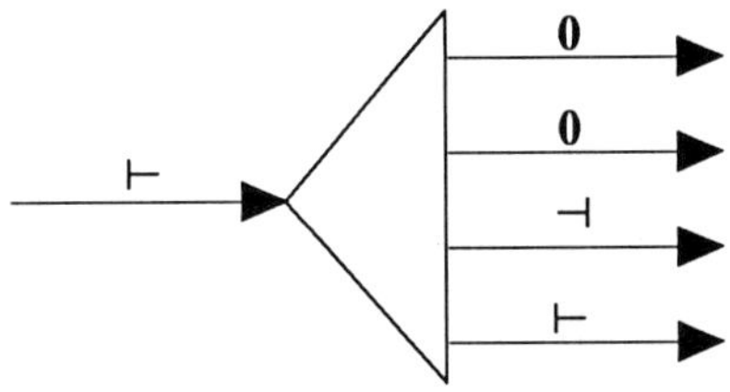

∨ :

Operator mode		Scalar mode
n = 1	n > 1	<x | ∨ | y>
∨ ∧ = ⊣	⊣	1
∨ \ = ⊢	⊢	1
∨ / = ∧	∧	1
∨ ↑ = \	**0**	0

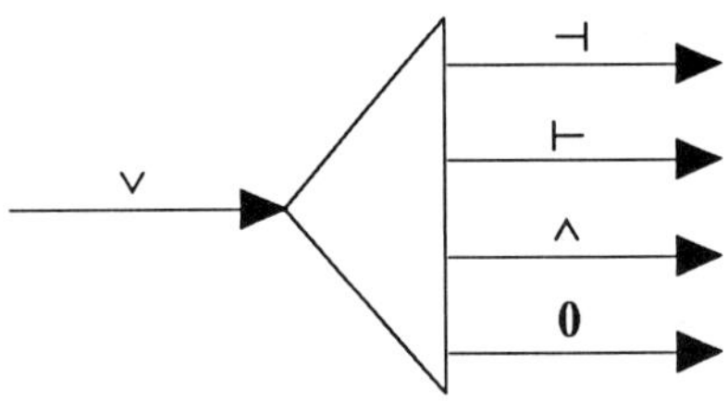

→ :

Operator mode		Scalar mode
n = 1	n > 1	<x | → | y>
→ ∧ = ⊣	⊣	1
→ \ = ⊢	⊢	1
→ / = /	**0**	0
→ ↑ = ↑	↑	1

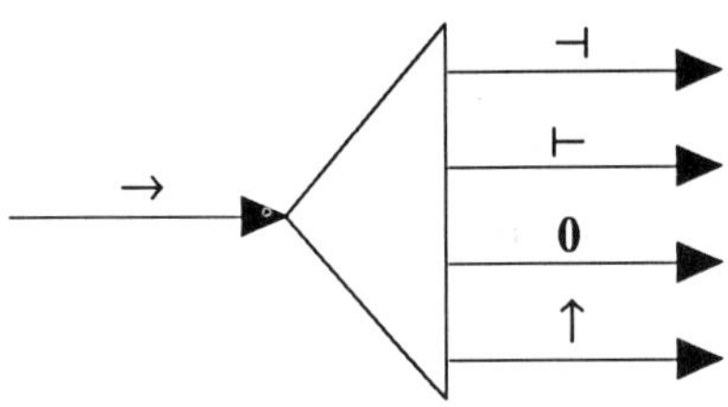

$\leftarrow$:

Operator mode		Scalar mode
n = 1	n > 1	$\langle x \vert \leftarrow \vert y \rangle$
$\leftarrow \wedge = \wedge$	$\wedge$	1
$\leftarrow \backslash = \backslash$	**0**	0
$\leftarrow / = \dashv$	$\dashv$	1
$\leftarrow \uparrow = \vdash$	$\vdash$	1

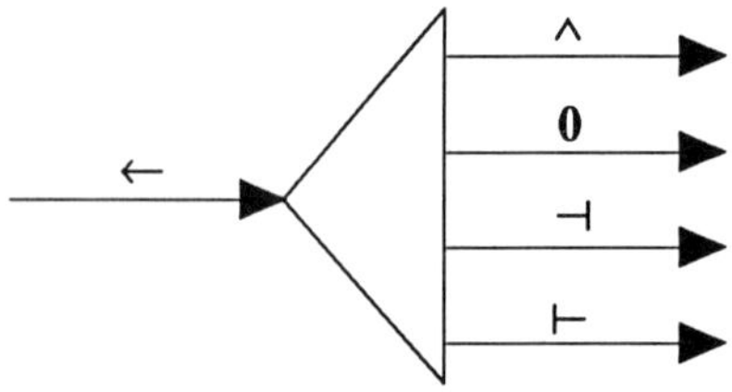

$\downarrow$:

Operator mode		Scalar mode
n = 1	n > 1	$\langle x \vert \downarrow \vert y \rangle$
$\downarrow \wedge = /$	**0**	0
$\downarrow \backslash = \uparrow$	$\uparrow$	1
$\downarrow / = \dashv$	$\dashv$	1
$\downarrow \uparrow = \vdash$	$\vdash$	1

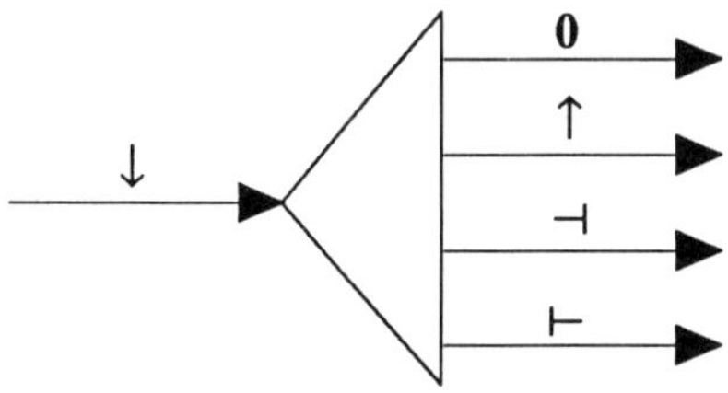

$\hat{1}$:

Operator mode		Scalar mode
n = 1	n > 1	$\langle x \vert \hat{1} \vert y \rangle$
$\hat{1} \wedge = \dashv$	$\dashv$	1
$\hat{1} \backslash = \vdash$	$\vdash$	1
$\hat{1} / = \dashv$	$\dashv$	1
$\hat{1} \uparrow = \vdash$	$\vdash$	1

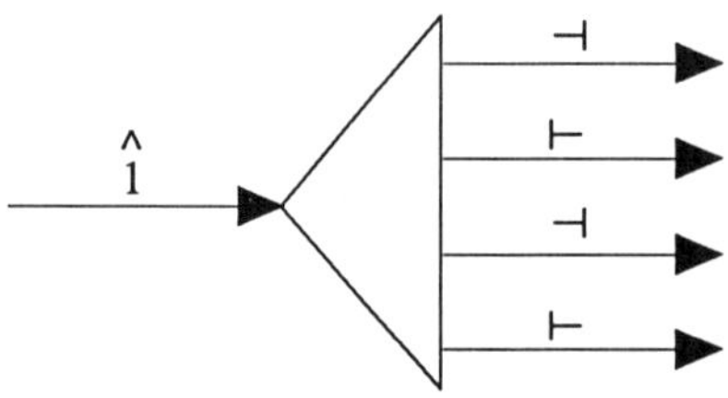

Autoproducts provide every logic function with a unique operator mapping or representation. The steady memory state is obtained in the first iteration, n = 2. For tautology $\hat{1}$ and for converse filters $\dashv$, $\vdash$ the steady states are already saturated in a single circle computation. The operator mode can serve as a standby memory, with no need for the halting instruction because for n > 1 the state of the memory remains unaltered.

Of particular interest are the autoproducts of the negative logic functions

$$- \langle \overset{\ulcorner \times \urcorner}{x \vert L} \vert y \rangle = -1^{(n)} \, (L \vert y \rangle \langle x \vert)^{n} ,$$

which are periodic, oscillating between positive and negative operators for even and odd values of n respectively.

The concept of an autoproduct, as developed above, allows us to consider autonomous logic with memory, where the scalar mode represents processing and the operator mode represents memory. Both functions, evaluation and memorization, are realizable on the same logic structure. The transition to the scalar processing mode via the operator mode of the autoproduct

$$\langle \overset{\ulcorner\ \times\ \urcorner}{x | L} | y\rangle = L | y\rangle \langle x | = | L y\rangle \langle x |$$

is obtained by transvection:

$$\mathrm{Tr}\, | L y\rangle \langle x | = \langle x | L | y\rangle .$$

Logic processing is a trace of memory. The trace is an immensely useful operation which establishes a direct dynamic link between memory and processing that is both natural and fundamental. Because autoproducts are at one and the same time units of memory and units of processing, modification of the memory state modifies accordingly the processing, and, vice versa, the inference process determines a certain pattern of memory. A simultaneous treatment of logic inference and memory, which includes the storing and retrieving of operators, is a significant departure from the classical theory where the characterization of a computer is accomplished through the study of its input. In defining autoproduct logic, we depart from the classical theory and define the machine as a *memory processor* with intrinsic autocontrol over its own operations.

3.3 Logic quantum numbers

There is a striking mathematical similarity between the formalisms of matrix logic and quantum theory. Both formalisms develop within vector spaces, logic space and Hilbert space respectively, and both are built around a similar mathematical construct. The bracket closure

$$\langle x | L | y\rangle ,$$

where $\langle x |$ and $| y\rangle$ are logic vectors from two adjoint vector spaces, and L a logic operator, is an atomic valuation device of logic, whereas, according to the Dirac formalism, a similar measuring device underlies quantum physics:

$$\langle \psi^* \mid Q \mid \psi \rangle \,,$$

where ψ^* and ψ are the state vectors or wave functions from two dual spaces, and Q a quantum-mechanical operator. In quantum physics an act of measurement, as described by the bracket closure $\langle \psi^* \mid Q \mid \psi \rangle$, projects a pure state out of the mixed state of a quantum system, whereas the logic closure can also extract a definite truth-value from a logic system which is capable of being in intermediary undecided states. As a matter of fact these can be viewed within two congruent frameworks: in the quantum domain as state vectors, and in the domain of logic as logic vectors.

Thus we are given two adjoint interpretations: the measurement act in the quantum domain parallels a decision process in the domain of logic. In accordance with the general scheme for translating between the two interpretations, what is seen from the logical view as an inference process can be described in quantum language as the collapse of the state vector.

In view of the links described above one cannot escape the impression that in matrix logic and in quantum physics we are dealing with different modes of one fundamental phenomenon. Could it be that matrix logic is actually revealing the logical processes to be quantum-physical in nature? Are there convincing arguments for considering the operations of the mind to be quantized? And if so, can we establish an explicit connection between logic and quantum physics? The matrix operator formulation of logic sheds new light on the fundamentals of logic and supports an important breakthrough in the study of the physics of the mind.

The new concept of logic quantum numbers, introduced in this section, indicates not only that the code of logic and the quantum code have the same fractal structure, but also that the problems of logic and of the intelligence code in general can be effectively formulated as eigenvalue problems similar to those of quantum physics.

3.3.1 Observables

Quantum theory fundamentally altered our understanding of reality and the way we describe it. The fact that the energy of quantum systems can obtain only discrete values

$$E_n = \hbar\omega\,(n + \tfrac{1}{2})\,, \qquad n = 0, 1, 2 \ldots$$

and the impossibility of ascribing an exact trajectory to a quantum particle gave rise to the operator formulation of quantum theory, where, in contrast with classical physics, observable quantities are represented not by continuous differentiable functions but by operators. On the atomic scale classical logic can no longer serve as a guide; for the description of quantized physical quantities, the continuous functions employed in classical physics are not applicable. Moreover, whereas in classical physics mathematical description represents physical phenomena directly, in quantum physics the physical states are represented by state vectors or wave functions which in themselves are not observable quantities but simply a mathematical device employed for the evaluation of probabilities.

The equations of quantum physics are quite similar to the equations of classical theory, and often it is possible to obtain quantum-mechanical equations simply by replacing observables in classical equations by suitable operators. The operators represent observable quantities and each observable quantity of a quantum system is associated with a Hermitian operator. In the act of observation of an operator we obtain the eigenvalues of the operator.

Generally, the eigenvalues can be found from the eigenequation

$$L\,|\,q\rangle = \lambda\,|\,q\rangle$$
$$(L - \lambda\, I)\,|\,q\rangle = 0$$

where L is an operator, $|\,q\rangle$ an eigenstate vector, and λ an eigenvalue.

The eigenequation possesses a non-null solution only when the resolvent matrix

$$(L - \lambda\, I)$$

is singular:

$$\det(L - \lambda\, I) = \begin{vmatrix} L_{11} - \lambda & L_{12} \\ L_{21} & L_{22} - \lambda \end{vmatrix} = 0$$

Solving the characteristic polynomial, we obtain the eigenvalues of the operator and then the corresponding eigenstates can be computed.

Before the operator description had fully replaced the classical methods, the equations for the eigenvalues of the operator of energy, the Hamiltonian, were mistakenly considered to be equations describing the propagation of real waves of matter, and not as equations that simply express the connection between operator and eigenvalues.

The application of an operator L to a state vector $|\,q\rangle$ is equivalent to expanding the vector by the factor λ. Thus the application of the operator reproduces the same vector multiplied by a scalar. Consequently, the orientation of the state vector remains unaltered. If a quantum system is in a state $|\,q\rangle$, which is an eigenstate of an operator L, and we observe a system in this state, then we are certain to get the corresponding eigenvalue λ. Conversely, when in the observation of an operator L we get the eigenvalue λ, then the quantum system is certain to be in the eigenstate $| q\rangle$, satisfying the eigenequation $L | q\rangle = \lambda | q\rangle$. If an eigenvalue admits several eigenvectors, the state of a system is degenerate.

There are two major classes of operators, one with a continuous spectrum of eigenvalues and the other with a discrete spectrum. Operators with the discrete spectrum play a crucial role in quantum theory, because they can be employed for the description of quantized quantum quantities.

Eigenstates associated with different eigenvalues of an operator are orthogonal. Because eigenstates can always be normalized, we have the orthonormality condition:

$$\langle q_i | q_j \rangle = \delta_{ij} .$$

Many relations involving operators are simplified if their eigenvectors are employed as the basis vectors, because the eigenvectors of an operator characterize the operator as such without reference to a particular basis system.

3.3.2 Logic eigenequations

The representation of observable quantities by operators is a unique feature of quantum theory which fundamentally distinguishes it from the rest of scientific inquiry. With the development of the operator logic theory of intelligence, quantum phenomena may no longer alone license the operator representation of observable quantities.

The cognition processes bear a strong resemblance to the phenomena we deal with in the quantum domain. The actual mechanism of reasoning is inaccessible. When we try to observe our own flow of thoughts, it quickly becomes clear that this observation is accompanied by unpredictable disturbances. Though it is possible to provide an approximate description of the thought process, an attempt to give an exact account introduces unpredictable disturbances, so that in the end we do not process the initial thought but think of how to describe it.

Just as quantum quantities cannot be measured with absolute precision, due to uncontrollable disturbances introduced by the act of observation, it is likewise not possible to pinpoint with absolute precision the underlying logic transformations of the thought process. The theory of logic quantum numbers leads to an important clarification of the intractability of the mechanism of thought and allows us for the first time to tackle the puzzle of high-level intelligence in a scientific manner.

In the footsteps of quantum physics we shall represent observable logic quantities by logic operators, associated with corresponding logic eigenstates and eigenvalues. In this fundamental approach, observation concerns not the classical logic variables but logic connective operators, that is, the logic operations themselves! The connectives, standing for logic operations, become the observable quantities, subject to mental registration. Moreover, in contrast to classical logic without magnitudes, logic quantities can now be characterized by measurable algebra.

The intelligent operations are described as logic operators acting upon logic states. The observation of a logic state $|q>$, associated with a logic operator L, yields the logic quantum number λ, which is an eigenvalue of the operator L.

The logic eigenequations can be seen as equations of real acts of the mind. However, they do not describe an actual cognitive transformation but simply a relation between logic observables and logic eigenstates.

If cognition is in an eigenstate, then the result of the observation of an operator is a logic quantum number, which characterizes the truth status of a logic system.

The logic operators parallel the Hamiltonian or energy operator H; logic quantum numbers parallel the quantized energy values E of a quantum system:

$$H\,|\psi> \;=\; E\,|\psi>\,.$$

Just as quantum numbers specify the eigenstates of a quantum system, the logic quantum numbers characterize the truth status of a logic system.

The results of this approach send a conceptual shock wave through the field of research into the mind. Whereas conventional logic operates with scalar values, considered to be directly observable, we describe logic truth states by logic vectors, which, being similar to a wave function, are not the observable entities but simply serve as a mathematical device for describing the logic states.

The observables in standard matrix logic are represented by (2×2) logic operators which can have not more than two eigenvalues.

The equation

$$L\,|q> \;=\; \lambda\,|q>\,,$$

where L is a logic operator and $|q>$ a logic vector, we shall call the logic eigenequation for obtaining the eigenvalues λ. All its non-null solutions $|q>$ we shall call logic eigenstate vectors of the operator L. The value λ for which there exists at least one logic eigenstate vector we call a logic eigenvalue of the operator L.

Eigenstates associated with different logic eigenvalues of an operator are linearly independent and orthogonal. Any logic vector can be expanded in the basis formed by the complete orthonormal set of eigenvectors.

As was previously indicated, eigenvalues are commonly obtained from the eigenequation by solving the characteristic polynomial

$$| L - I\lambda | = 0 .$$

However, this is not the only way to obtain eigenvalues. Other considerations may also be brought to bear.

Identity and empty operator

There is no need to solve the characteristic polynomial in the case of the identity operator I and the empty operator $\hat{0}$. Because these logic operators represent the matrix equivalent of the numbers 1 and 0 respectively, the logic eigenvalues are obtained directly:

$$I | q> = \lambda | q> \quad \Rightarrow \quad \lambda = 1$$
$$\hat{0} | q> = \lambda | q> \quad \Rightarrow \quad \lambda = 0 .$$

Clearly, since any logic state vectors satisfy these eigenequations, the logic eigenstates in both cases are degenerate.

Conjunction and NOR

We can compute the eigenvalues for the projective operators of conjunction $\wedge$ and NOR, taking into account that these are idempotent logic operators. Due to idempotency it follows that

$$L^2 | q> = L | q> ,$$

that is,

$$\lambda^2 | q> = \lambda | q> .$$

Therefore

$$\lambda^2 = \lambda$$
$$\lambda^2 - \lambda = 0$$
$$\lambda(\lambda - 1) = 0$$
$$\lambda_1 = 0, \qquad \lambda_2 = 1$$

The logic law of contradiction applies for eigenvalues of idempotent logic operators. The eigenstates associated with the eigenvalues 0 and 1 are:

— for conjunction $\wedge$,

$$|q_1\rangle = \begin{pmatrix}1\\0\end{pmatrix}, \qquad \lambda_1 = 0$$

$$|q_2\rangle = \begin{pmatrix}0\\1\end{pmatrix}, \qquad \lambda_2 = 1$$

— for NOR operator $\uparrow$,

$$|q_1\rangle = \begin{pmatrix}1\\0\end{pmatrix}, \qquad \lambda_1 = 1$$

$$|q_2\rangle = \begin{pmatrix}0\\1\end{pmatrix}, \qquad \lambda_2 = 0$$

Negation operator

The calculation of eigenvalues of the negation operator is also straightforward: since the square of negation is identity

$$\neg^2 = I,$$

we derive

$$\neg^2 |q\rangle = \lambda^2 |q\rangle = |q\rangle .$$

Thus

$$\lambda^2 = 1$$

$$\lambda_1 = 1, \qquad \lambda_2 = -1$$

Substituting $\lambda_{1,2} = \pm 1$ into the eigenequation for the negation operator

$$\neg |q\rangle = \lambda |q\rangle ,$$

we get:

$$\begin{cases} -\bar{q}_1 + q_1 = 0 \\ \bar{q}_1 - q_1 = 0 \end{cases}$$

$$\begin{cases} \bar{q}_2 + q_2 = 0 \\ \bar{q}_2 + q_2 = 0 \,. \end{cases}$$

Taking into account the general normalization condition

$$|\bar{q}| + |q| = 1 \,,$$

we obtain two eigenstate vectors of the negation operator:

$$|q_1> \;=\; \begin{pmatrix} \frac{1}{2} \\ \frac{1}{2} \end{pmatrix}, \qquad \lambda_1 = 1$$

$$|q_2> \;=\; \begin{pmatrix} \frac{1}{2} \\ -\frac{1}{2} \end{pmatrix}, \qquad \lambda_2 = -1 \,.$$

<u>Tautology operator</u>

The eigenstates of the tautology operator $\hat{1}$ coincide with the eigenstates of the negation operator, but the logic quantum numbers associated with these states are different:

$$|q_1> \;=\; \begin{pmatrix} \frac{1}{2} \\ \frac{1}{2} \end{pmatrix}, \qquad \lambda_1 = 2$$

$$|q_2> \;=\; \begin{pmatrix} \frac{1}{2} \\ -\frac{1}{2} \end{pmatrix}, \qquad \lambda_2 = 0 \,.$$

Whereas negation switches over the falsum and verum components of the logic eigenstate:

$$\neg \begin{pmatrix} \frac{1}{2} \\ \frac{1}{2} \end{pmatrix} = 1 \begin{pmatrix} \frac{1}{2} \\ \frac{1}{2} \end{pmatrix} = \begin{pmatrix} \frac{1}{2} \\ \frac{1}{2} \end{pmatrix}$$

and

$$\neg\begin{pmatrix}\frac{1}{2}\\-\frac{1}{2}\end{pmatrix} = -1\begin{pmatrix}\frac{1}{2}\\-\frac{1}{2}\end{pmatrix} = \begin{pmatrix}-\frac{1}{2}\\\frac{1}{2}\end{pmatrix},$$

the tautology operator acts on the eigenvectors quite differently:

$$\hat{1}\begin{pmatrix}\frac{1}{2}\\\frac{1}{2}\end{pmatrix} = 2\begin{pmatrix}\frac{1}{2}\\\frac{1}{2}\end{pmatrix} = \begin{pmatrix}1\\1\end{pmatrix}$$

and

$$\hat{1}\begin{pmatrix}\frac{1}{2}\\-\frac{1}{2}\end{pmatrix} = 0\begin{pmatrix}\frac{1}{2}\\-\frac{1}{2}\end{pmatrix} = \begin{pmatrix}0\\0\end{pmatrix}.$$

Here the falsum and verum components of the eigenstate vectors are summed. When the components are in the parallel mode of the same sign, this is equivalent to multiplying the eigenstate by the logic quantum number 2. When the components are in the antiparallel mode of opposite signs, this is equivalent to multiplying the eigenstate by the logic quantum number 0. The coupling of falsum and verum in the antiparallel mode of tautology resembles the coupling of two fermion states into a boson state in the superconductivity model.

Disjunction and NAND

Considering the eigenequations for the disjunction operator $\vee$ and NAND $\downarrow$,

$$\vee\,|q\rangle = \lambda\,|q\rangle$$
$$\downarrow|q\rangle = \lambda\,|q\rangle$$

we establish that the only logic vector satisfying the equations is the undecided null vector:

$$|\mathbf{0}\rangle = \begin{pmatrix}0\\0\end{pmatrix}.$$

Because a null vector cannot be an eigenstate solution of an eigenequation, we must conclude that in real logic space the disjunction operator and the NAND operator are not observable quantities! The λ in the equation can take arbitrary values, which implies that no eigenstates can be assigned to the operators $\vee$ and $\downarrow$. This fundamentally distinguishes $\vee$ and $\downarrow$ from the other logic operators, and indicates a principal difference between the minor basis sets

$$\{\urcorner, \wedge\} \quad \text{and} \quad \{\urcorner, \vee\} .$$

In conventional logic the minor basis sets $\{\urcorner, \wedge\}$ and $\{\urcorner, \vee\}$ are considered to be fully equivalent: employing the duality transformation, it is always possible to pass from one basis set to another. However, matrix logic reveals a fundamental difference between the operators $\wedge$ and $\vee$. Whereas the operator $\wedge$ is an observable quantity with definite eigenstates and eigenvalues, the disjunction operator $\vee$ by contrast does not have eigenvectors satisfying the eigenequation, except for the null vector, which cannot be a solution. It now becomes apparent that the classical minor basis sets are physically not equivalent! Recall that in Boolean algebra, disjunction is expressed as logic addition, including the physically inadmissible operation:

$$1 + 1 = 1 .$$

Disjunction $\vee$ is not capable of preserving invariant orientation in logic space, and therefore does not represent an observable operator. The operators $\vee$ and $\downarrow$ are not observable in real physical conditions, only in artificial logic space, such as provided by the computer.

In the chart below we collect the eigenvalues and the corresponding eigenstate vectors for the strictly binary subset of logic quantum numbers. This includes both the left eigenequation,

$$L\,|\,q> = \lambda\,|\,q> ,$$

and the right eigenequation,

$$<q\,|\,L = <q\,|\,\lambda .$$

The diagonal symmetric logic operators $\wedge$ and $\uparrow$ have two eigenvalues for both left and right eigenstates. The contrasymmetric operators $\rightarrow, \leftarrow, \backslash, /$ have just one left and one right eigenvalue.

Logic Operator	Eigenvalue(s)		Eigenvector(s)	
	Left	Right	Left	Right
$\hat{0}$	0	0	any	any
I	1	1	any	any
$\wedge$	1, 0	1, 0	$\vert 1\rangle$, $\vert 0\rangle$	$\langle 1 \vert$, $\langle 0 \vert$
$\uparrow$	0, 1	0,1	$\vert 1\rangle$, $\vert 0\rangle$	$\langle 1 \vert$, $\langle 0 \vert$
$\rightarrow$	1	1	$\vert 0\rangle$	$\langle 1 \vert$
$\leftarrow$	1	1	$\vert 1\rangle$	$\langle 0 \vert$
\	0	0	$\vert 1\rangle$	$\langle 0 \vert$
/	0	0	$\vert 0\rangle$	$\langle 1 \vert$

The observable operator set of filters

$$\{\perp, \mathrm{T}, \dashv, \vdash\}$$

occupies an intermediary position between the strictly binary subset of logic eigenstates and the Fermi states that characterize the fundamental operators of negation ⌉ and tautology $\hat{1}$. The eigenvalues of filters are the binary logic numbers 0 and 1, but eigenvectors may include both binary and fractional logic vectors. The following chart offers an explicit presentation of the eigenequations for filters:

Left logic	Right logic
$\begin{pmatrix}0&0\\1&1\end{pmatrix}\begin{pmatrix}\frac{1}{2}\\-\frac{1}{2}\end{pmatrix} = 0\begin{pmatrix}\frac{1}{2}\\-\frac{1}{2}\end{pmatrix} = \begin{pmatrix}0\\0\end{pmatrix}$ $\begin{pmatrix}0&0\\1&1\end{pmatrix}\begin{pmatrix}0\\1\end{pmatrix} = 1\begin{pmatrix}0\\1\end{pmatrix} = \begin{pmatrix}0\\1\end{pmatrix}$	$(1\ \ 0)\begin{pmatrix}0&0\\1&1\end{pmatrix} = (1\ \ 0)\,0 = (0\ \ 0)$ $(\frac{1}{2}\ \ \frac{1}{2})\begin{pmatrix}0&0\\1&1\end{pmatrix} = (\frac{1}{2}\ \ \frac{1}{2})\,1 = (\frac{1}{2}\ \ \frac{1}{2})$
$\begin{pmatrix}1&1\\0&0\end{pmatrix}\begin{pmatrix}1\\0\end{pmatrix} = 1\begin{pmatrix}1\\0\end{pmatrix} = \begin{pmatrix}1\\0\end{pmatrix}$ $\begin{pmatrix}1&1\\0&0\end{pmatrix}\begin{pmatrix}\frac{1}{2}\\-\frac{1}{2}\end{pmatrix} = 0\begin{pmatrix}\frac{1}{2}\\-\frac{1}{2}\end{pmatrix} = \begin{pmatrix}0\\0\end{pmatrix}$	$(\frac{1}{2}\ \ \frac{1}{2})\begin{pmatrix}1&1\\0&0\end{pmatrix} = (\frac{1}{2}\ \ \frac{1}{2})\,1 = (\frac{1}{2}\ \ \frac{1}{2})$ $(0\ \ 1)\begin{pmatrix}1&1\\0&0\end{pmatrix} = (0\ \ 1)\,0 = (0\ \ 0)$
$\begin{pmatrix}0&1\\0&1\end{pmatrix}\begin{pmatrix}1\\0\end{pmatrix} = 0\begin{pmatrix}1\\0\end{pmatrix} = \begin{pmatrix}0\\0\end{pmatrix}$ $\begin{pmatrix}0&1\\0&1\end{pmatrix}\begin{pmatrix}\frac{1}{2}\\\frac{1}{2}\end{pmatrix} = 1\begin{pmatrix}\frac{1}{2}\\\frac{1}{2}\end{pmatrix} = \begin{pmatrix}\frac{1}{2}\\\frac{1}{2}\end{pmatrix}$	$(\frac{1}{2}\ \ -\frac{1}{2})\begin{pmatrix}0&1\\0&1\end{pmatrix} = (\frac{1}{2}\ \ -\frac{1}{2})\,0 = (0\ \ 0)$ $(0\ \ 1)\begin{pmatrix}0&1\\0&1\end{pmatrix} = (0\ \ 1)\,1 = (0\ \ 1)$
$\begin{pmatrix}1&0\\1&0\end{pmatrix}\begin{pmatrix}\frac{1}{2}\\\frac{1}{2}\end{pmatrix} = 1\begin{pmatrix}\frac{1}{2}\\\frac{1}{2}\end{pmatrix} = \begin{pmatrix}\frac{1}{2}\\\frac{1}{2}\end{pmatrix}$ $\begin{pmatrix}1&0\\1&0\end{pmatrix}\begin{pmatrix}0\\1\end{pmatrix} = 0\begin{pmatrix}0\\1\end{pmatrix} = \begin{pmatrix}0\\0\end{pmatrix}$	$(1\ \ 0)\begin{pmatrix}1&0\\1&0\end{pmatrix} = (1\ \ 0)\,1 = (1\ \ 0)$ $(\frac{1}{2}\ \ -\frac{1}{2})\begin{pmatrix}1&0\\1&0\end{pmatrix} = (\frac{1}{2}\ \ -\frac{1}{2})\,0 = (0\ \ 0)$

We can distinguish four kinds of logic operators:

(1) The set of 12 logic operators

$$\{I, \wedge, \uparrow, \rightarrow, \leftarrow, \backslash, /, \perp, T, \dashv, \vdash, \hat{0}\}$$

whose eigenvalues belong to the binary set $E_2 = \{0,1\}$.

(2) The fundamental logic operator $\rceil$, whose logic eigenvalues include the negative logic value –1,

$$\lambda_1 = 1, \qquad \lambda_2 = -1 .$$

(3) The tautology operator $\hat{1}$, whose eigenvalues include a non-binary value 2,

$$\lambda_1 = 2, \qquad \lambda_2 = 0 .$$

(4) The operators of disjunction $\vee$ and NAND $\downarrow$, which can be characterized by arbitrary eigenvalues, and in this sense are not observable!

We have established four kinds of eigenstate logic vectors:

(1) The operator set $\{\wedge, \uparrow, \rightarrow, \leftarrow, \backslash, /\}$ has binary eigenstate vectors that are base logic vectors:

$$(0,1) \quad \text{and} \quad (1, 0) .$$

(2) The eigenstate vectors for the operator set $\{\rceil, \hat{1}\}$ include negative logic values.

(3) The operators $\{\perp, \mathrm{T}, \dashv, \vdash\}$ can have eigenstate vectors from both classes.

(4) The operators I and $\hat{0}$ can have arbitrary eigenstate vectors. By comparison, the unobservable operators $\vee$ and $\downarrow$ can be ascribed arbitrary logic numbers.

Particular importance attaches to the observable logic operators with complete and orthogonal sets of eigenstates. This concerns only four logic observables:

$$\wedge: \quad \langle q_1 | q_2 \rangle = (1 \;\; 0)\begin{pmatrix} 0 \\ 1 \end{pmatrix} = 0$$

$$\uparrow: \quad \langle q_1 | q_2 \rangle = (0 \;\; 1)\begin{pmatrix} 1 \\ 0 \end{pmatrix} = 0$$

$$\rceil: \quad \langle q_1 | q_2 \rangle = \left(\tfrac{1}{2} \;\; \tfrac{1}{2}\right)\begin{pmatrix} \tfrac{1}{2} \\ -\tfrac{1}{2} \end{pmatrix} = 0$$

$$\hat{1}: \quad \langle q_1 | q_2 \rangle = \left(\tfrac{1}{2} \;\; \tfrac{1}{2}\right)\begin{pmatrix} \tfrac{1}{2} \\ -\tfrac{1}{2} \end{pmatrix} = 0 .$$

Though there are also two eigenstates for the filter operators

$$\{\perp, \top, \dashv, \vdash\},$$

in comparison with $\{\wedge, \uparrow, \rceil, \hat{1}\}$ their eigenstates neither form a vector space basis nor are orthogonal. This has fundamental consequences. Generally, a logic state can be expanded only in the orthonormal bases of the eigenstates of conjunction $\wedge$, NOR $\uparrow$, negation $\rceil$ and tautology $\hat{1}$.

The eigenstates of the projective operators $\wedge$ and $\uparrow$ allow us to accommodate only the binary logic eigenvalues:

$$\lambda \in E_2, \qquad E_2 = \{0, 1\}.$$

The observables $\rceil$ and $\hat{1}$ can generate the complete set of logic quantum numbers

$$\lambda \in E_4, \qquad E_4 = \{-1, 0, 1, 2\},$$

while an arbitrary logic vector can be extended in their orthonormal bases. In this sense, the fundamental set of observables $\{\rceil, \hat{1}\}$ is sufficient to deal with any logic problem!

An important property of the logic eigenstates for $\wedge$ and $\uparrow$ is that an eigenstate associated with the eigenvalue 1 can be obtained by complementation of an eigenstate associated with the eigenvalue 0. Here complementation of states corresponds to complementation of logic quantum numbers. This is a unique property of the eigenstates of the projective logic operators, where the complement operator can be employed as an inverter or switcher of the eigenstates. Such switching is not applicable to the eigenstates of either $\{\rceil, \hat{1}\}$ or $\{\perp, \top, \dashv, \vdash\}$.

The investigation of logic quantum numbers reveals that the eigenvalues of logic operators take only discrete values:

$$\lambda_i = \{-1, 0, 1, 2\}.$$

This is not only an indication that the intelligence code is quantized but, what is more, the eigenvalue spectrum of logic operators coincides exactly with the fundamental alphabet of logic truth-values,

$$E_4 = \{-1, 0, 1, 2\},$$

which was obtained on completely different grounds!

3.3.3 Compatibility

Because matrix logic operations are generally non-commutative, the fundamental commutation limit applies to the process of logic evaluations and observation. This brings into focus the important question of the compatibility of logic observables.

A simultaneous observation of two logic quantities requires commutation of their respective operators:

$$[L_1, L_2] = L_1L_2 - L_2L_1 = 0 .$$

Compliance with the commutation relation is a rare commodity in matrix logic. Of course, every operator L commutes with itself, with identity and with the empty operator:

$$[L, L] = 0$$
$$[L, I] = 0$$
$$[L, \hat{0}] = 0 ,$$

but these trivial cases are of little interest to us. An important fact is that most of the logic operators do not commute, and a number of commutation products violate the binarity condition. However, there are fundamental sets of commuting logic operators, precisely those operators which are absolutely essential in logic evaluations. The fundamental condition for commutation of logic observables requires that they can share the same eigenstate. Only then are the logic operators compatible and can simultaneous observation become possible.

We are rewarded with the following fundamental set of commuting logic operators:

$$[\wedge, \uparrow] = 0$$
$$[\rightarrow, /] = 0$$
$$[\leftarrow, \backslash] = 0$$
$$[\hat{1}, \rceil] = 0$$

In addition to those operators we have already analysed in detail we have, among the compatible operators, two pairs of dual operators: implication commutes with converse inhibition:

$$[\rightarrow, /] = / - / = 0 ,$$

and converse implication commutes with inhibition:

$$[\leftarrow, \backslash] = \backslash - \backslash = 0 .$$

The observables $\rightarrow$ and $/$ share the same eigenstates:

$$|0> \quad - \quad \text{for left logic}$$
$$<1| \quad - \quad \text{for right logic}\ .$$

Similarly, the observables $\leftarrow$ and $\backslash$:

$$|1> \quad - \quad \text{for left logic}$$
$$<0| \quad - \quad \text{for right logic}\ .$$

A principal difference between implication and inhibition and the remaining set of commuting operators is that the former have a single eigenstate. In left logic, implication does not alter the false eigenvector (consequent):

$$\rightarrow|0> = 1|0> = |0>\ ,$$

and converse inhibition nullifies the false eigenvector:

$$/|0> = 0|0> = |\mathbf{0}>\ .$$

In right logic, implication reproduces the true eigenvector (antecedent):

$$<1|\rightarrow = <1|1 = <1|\ ,$$

and converse inhibition nullifies the true eigenvector:

$$<1|/ = <1|0 = <\mathbf{0}|\ .$$

The commutation of the projective operators $\wedge$ and $\uparrow$ is rather obvious not only because both share the same logic eigenstates, but also because their products are empty:

$$\wedge\uparrow = \hat{0}\ , \qquad \uparrow\wedge = \hat{0}\ ,$$

and the commutation follows naturally:

$$[\wedge, \uparrow] = \hat{0} - \hat{0} = \hat{0}\ .$$

The most important cross-commutation occurs in the fundamental set of operators comprising tautology, negation and identity:

$$[\hat{1}, \neg] = 0$$
$$[\hat{1}, I] = 0$$
$$[\neg, I] = 0 .$$

The tautology operator is observable simultaneously with the negation operator and with the affirmation operator. This is in full agreement with the characteristic property of tautologies that are valid under any interpretation. The fundamental feature of the intelligence code is reaffirmed in terms of commutation.

3.3.4 Connectives as numbers

If the logic space of a problem comprises nothing but eigenstates, the logic operators can be replaced by the logic quantum numbers

$$\lambda_i = \{-1, 0, 1, 2\} .$$

Consequently, a matrix logic expression can be converted into a numerical form from which logic operators as such are completely eliminated.

We may recall that in vector logic we already set a precedent of replacing operators by their respective outer products. However, the elimination of operators by replacing them by logic quantum numbers is much more than a mathematical device, and is a step with fundamental scientific repercussions.

Clearly, the replacement of an operator by a logic number is possible when an associated logic vector is the eigenvector of that operator. Consider an atomic device:

$$\langle 0 | \wedge | 1 \rangle = 0 .$$

With respect to the ket eigenstate $| 1\rangle$ the conjunction can be replaced by 1; with respect to bra eigenstate $\langle 0 |$ it can be replaced by 0. Consequently,

$$\langle 0 | \wedge | 1 \rangle = \langle 0 | 1 | 1 \rangle = \langle 0 | 0 | 1 \rangle = 0 .$$

The replacement of the operator by the corresponding logic number does not in either case alter the value of the conjunction bracket product.

In the case under discussion both the ket and the bra logic vectors are accepted as the eigenstate vectors for conjunction. In fact, a single eigenstate is sufficient to eliminate an operator in an atomic product. Consider the elimination of the tautology operator:

$$<p|\hat{1}|q_1> \;=\; <p|2|\begin{matrix}\frac{1}{2}\\ \frac{1}{2}\end{matrix}> \;=\; 1$$

$$<p|\hat{1}|q_2> \;=\; <p|0|\begin{matrix}\frac{1}{2}\\ -\frac{1}{2}\end{matrix}> \;=\; 0\,.$$

The results of evaluation will stand for an arbitrary, in the first case, logically normalized, vector $<p|$.

If a logic vector is not the eigenvector of an operator with a complete orthonormal set of eigenstates, we must first expand the logic vector in the basis of eigenvectors. Consider

$$\neg\begin{pmatrix}1\\0\end{pmatrix} = \begin{pmatrix}0\\1\end{pmatrix}.$$

The elimination of the negation operator cannot be done directly, since $|0>$ is not an eigenvector of the operator. However, expanding $|0>$ in the basis of the eigenvectors of negation, we obtain:

$$\neg\begin{pmatrix}1\\0\end{pmatrix} = \neg\left[\begin{pmatrix}\frac{1}{2}\\ \frac{1}{2}\end{pmatrix} + \begin{pmatrix}\frac{1}{2}\\ -\frac{1}{2}\end{pmatrix}\right].$$

Due to the linearity of matrix operations:

$$\neg\begin{pmatrix}1\\0\end{pmatrix} = \neg\begin{pmatrix}\frac{1}{2}\\ \frac{1}{2}\end{pmatrix} + \neg\begin{pmatrix}\frac{1}{2}\\ -\frac{1}{2}\end{pmatrix}.$$

Now the negation operator acts upon the eigenvectors and can be replaced by the corresponding eigenvalues:

$$\neg\begin{pmatrix}1\\0\end{pmatrix} = 1\begin{pmatrix}\frac{1}{2}\\ \frac{1}{2}\end{pmatrix} + (-1)\begin{pmatrix}\frac{1}{2}\\ -\frac{1}{2}\end{pmatrix} = \begin{pmatrix}0\\1\end{pmatrix}.$$

The identification of the logic operators with logic observables, and the possibility of replacing of an operator L by a corresponding eigenvalue λ, in accordance with the eigenequation

$$L\,|\,q> \;=\; \lambda\,|\,q>\,,$$

lead to a new fundamental formulation of logic theory. Reducing a logic expression to the canonical form of a fundamental product, comprising summands,

$$<x_1\,|\,L_1\,|\,y_1><x_2\,|\,L_2\,|\,y_2> \ldots <x_n\,|\,L_n\,|\,y_n>\,,$$

we can replace the string of logic operators

$$L_1\,|\,y_1><x_2\,|\,L_2\,|\,y_2> \ldots <x_n\,|\,L_n$$

with the corresponding sequence of eigenvalues

$$<x_1\,|\,\lambda_1\,\lambda_2\,\lambda_3 \ldots \lambda_{2n-1}\,|\,y_n>\,,$$

provided that the eigenstate conditions are satisfied.

The number of eigenvalues in the product can be reduced to a single factor, if the multiplication of operators is carried out. Consider an operator–vector product

$$L_1\,L_2\,|\,q> \;=\; L_3\,|\,q>\,,$$

where $|\,q>$ is an eigenstate vector for L_1, L_2 and $L_3 = L_1\,L_2$.The operators L_1, L_2 can be replaced by the corresponding eigenvalues in the following way:

$$L_1\,L_2\,|\,q> \;=\; L_1\,\lambda_2\,|\,q> \;=\; \lambda_2\,L_1\,|\,q> \;=\; \lambda_2\,\lambda_1\,|\,q>\,,$$

and

$$L_3\,|\,q> \;=\; \lambda_3\,|\,q>\,.$$

Consequently, the eigenvalue of an operator product is a product of eigenvalues:

$$\lambda_3 \;=\; \lambda_1\,\lambda_2\,.$$

The number multiplication code for logic connective operators corresponding to the multiplication table of matrix logic laid out above is presented below:

×	**0**	I	→	←	\	/	∧	↑	⅂	**1**	⊥	T	⊣	⊢
0	0	0	0	0	0	0	0	0	0	0	0	0	0	0
I	0	1	1	1	0	0	0 / 1	1 / 0	1 / −1	2 / 0	0 / 1	1 / 0	0 / 1	1 / 0
→	0	1	1			0	0 /	1 /				1 /	0 /	
←	0	1		1	0		/ 1	/ 0			/ 1			/ 0
\	0	0		0	0		/ 0	0 /			/ 0			/ 0
/	0	0	0			0	0 /	0 /				0 /	0 /	
∧	0	0 / 1	0 /	/ 1	/ 0	0 /	0 / 1	0 / 0			/ 1	0 /	0 /	/ 0
↑	0	1 / 0	1 /	/ 0	0 /	0 /	0 / 0	1 / 0			/ 0	1 /	0 /	/ 0
⅂	0	1 / −1							1 / 1	2 / 0	0	0	1	1
1	0	2 / 0							2 / 0	2 / 0	0	0	2	2
⊥	0	0 / 1		/ 1	/ 0		/ 1	/ 0	0	0	0 / 1	0		0
T	0	1 / 0	1 /			0 /	0 /	1 /	0	0	0	1 / 0	0	
⊣	0	0 / 1	0 /			0 /	0 /	0 /	1	2		0	0 / 1	1
⊢	0	1 / 0		/ 0	/ 0		/ 0	/ 0	1	2	0		1	1 / 0

Table 3.6 Number multiplication code for logic connectives

The operators ∨ and ↓, which do not possess eigenstate vectors, are excluded from the number multiplication code, as are the interactions which cannot share the same eigenstate. The number multiplication code provides an interesting way of resolving the problem of forbidden interactions. Consider the product

$$<1\,|\rightarrow\rightarrow|\,0>\,,$$

containing the forbidden interaction $\rightarrow \rightarrow$. Because the logic vectors $|0\rangle$ and $\langle 1|$ are the right and left eigenvectors of the implication operator, the latter can be replaced by its eigenvalue:

$$\langle 1 | 1 \cdot 1 | 0 \rangle = \langle 1 | 0 \rangle .$$

No logic values inadmissible in discrete logic appear in the result.

It might be useful to trace back our line of reasoning. Starting from classical *symbolic* connectives which in no way reflect rational procedures involving number or quantity, we have moved to Boolean logic algebra, which identifies the conjunction and disjunction connectives with the *operations* of multiplication and addition respectively. We then made a major leap forward by interpreting logic connectives as matrix operators, which enabled us to formalize the complete set of logic connectives. Finally, the operator formulation of logic was brought to completion with the fundamental concept of logic quantum numbers:

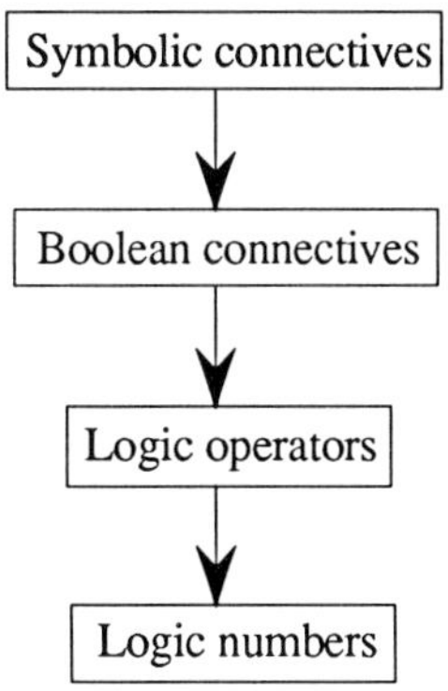

The possibility of representing logic connectives by numbers has fundamental consequences for logic theory, its application and limitations. A major obstacle on the path of the scientific study of high-level intelligence has been removed.

3.3.5 Diagonalization

The system of eigenequations for logic observables,

$$\begin{cases} L|q_1\rangle = \lambda_1 |q_1\rangle \\ L|q_2\rangle = \lambda_2 |q_2\rangle \end{cases}$$

can be joined into a single matrix equation:

$$L(|q_1> \ \ |q_2>) = (|q_1> \ \ |q_2>)\begin{pmatrix}\lambda_1 & 0\\ 0 & \lambda_2\end{pmatrix},$$

where L is a logic operator, $(|q_1> \ |q_2>)$ a juxtaposition matrix whose columns are the eigenvectors of the operator L, and λ_1, λ_2 its eigenvalues.
This matrix equation allows us to diagonalize the logic operator as follows:

$$(|q_1> \ \ |q_2>)^{-1} L(|q_1> \ \ |q_2>) = \begin{pmatrix}\lambda_1 & 0\\ 0 & \lambda_2\end{pmatrix},$$

where $(|q_1> \ |q_2>)^{-1}$ is the matrix inverse of $(|q_1> \ |q_2>)$.
The logic observables $\wedge$ and $\uparrow$ do not require diagonalization since they are already diagonal operators, with the diagonal comprising their eigenvalues. With regard to the diagonalization of the fundamental operators of negation and tautology, it is necessary to emphasize that the matrices

$$(|q_1> \ |q_2>)^{-1} \qquad \text{and} \qquad (|q_1> \ |q_2>)$$

involved in their diagonalization are themselves the logic operators of the tautology series and the corresponding fractional inverses respectively:

Negation

$$(|q_1> \ \ |q_2>)^{-1} = \begin{pmatrix}1 & 1\\ 1 & -1\end{pmatrix}$$

$$(|q_1> \ \ |q_2>) = \begin{pmatrix}\frac{1}{2} & \frac{1}{2}\\ \frac{1}{2} & -\frac{1}{2}\end{pmatrix} = \begin{pmatrix}1 & 1\\ 1 & -1\end{pmatrix}^{-1}$$

Tautology

$$(|q_1> \ \ |q_2>)^{-1} = \begin{pmatrix}1 & -1\\ 1 & 1\end{pmatrix}$$

$$(|q_1> \ \ |q_2>) = \begin{pmatrix}\frac{1}{2} & \frac{1}{2}\\ -\frac{1}{2} & \frac{1}{2}\end{pmatrix} = \begin{pmatrix}1 & -1\\ 1 & 1\end{pmatrix}^{-1}$$

The diagonalization equations are as follows:

Negation

$$\begin{pmatrix}1 & 1\\ 1 & -1\end{pmatrix}\begin{pmatrix}0 & 1\\ 1 & 0\end{pmatrix}\begin{pmatrix}\frac{1}{2} & \frac{1}{2}\\ \frac{1}{2} & -\frac{1}{2}\end{pmatrix} = \begin{pmatrix}1 & 0\\ 0 & -1\end{pmatrix}$$

Tautology

$$\begin{pmatrix}1 & -1\\ 1 & 1\end{pmatrix}\begin{pmatrix}1 & 1\\ 1 & 1\end{pmatrix}\begin{pmatrix}\frac{1}{2} & \frac{1}{2}\\ -\frac{1}{2} & \frac{1}{2}\end{pmatrix} = \begin{pmatrix}0 & 0\\ 0 & 2\end{pmatrix}$$

Because the determinant of a diagonal matrix is a product of diagonal elements, a matrix operator is nonsingular if and only if all its eigenvalues are different from zero. Clearly, from the diagonal form of negation and tautology it follows that the first is a nonsingular and the second a singular operator.

The fundamental importance of the diagonalization of logic operators is reflected in the fact that the logic observables comprise (a) a decuplet of diagonalizable operators or operators already diagonalized:

$$\{\urcorner, \hat{1}, \bot, \top, \dashv, \vdash, \wedge, \uparrow, \mathrm{I}, \hat{0}\}$$

and (b) a quadruplet of undiagonalizable operators:

$$\{\rightarrow, \leftarrow, \backslash, /\}$$

Because only operators with a complete orthogonal set of eigenvalues can be diagonalized, this does not apply to the quadruplet of asymmetric logic operators, which each possess just a single eigenvalue and a single eigenstate.

3.3.6 Fractal quantum mechanics

Niels Bohr, the founder of the modern theory of the atom, rather simplistically hypothesized that the cognition process involves such negligible amounts of energy that quantum-theoretical limitations may play an essential role in determining its character. The reflexive symmetry of quantum physics and matrix operator logic reinforces the position that the intelligence code must be formulated within the framework of the general covariant laws of physics. However, the answer to the question of whether the operations of the mind are quantum-mechanical in nature depends on whether the discrete code of high-level intelligence can be equated with or simply parallels the quantum-physical quantization. Though the non-commuting algebra of the mind possesses many fundamental features of quantum phenomena, it is not identical to these. It is necessary to emphasize that matrix logic is not a quantum logic which simply seeks to satisfy the paradoxes of the quantum world, but a formalism which establishes a quantum code of logic itself. Though this code resembles the rules and principles of quantum mechanics, it is at the same time qualitatively different.

Whereas in quantum physics the state vectors or wave functions serve as a formal instrument for the evaluation of probabilities, in matrix logic the truth-value components of logic vectors are already interpretable as probabilities, characterizing different degrees of validity. This indicates a principal difference between the logic vectors participating in logic equations and the wave functions used in quantum mechanics. The intricate mechanics of the mind must be viewed as a phenomenon of a higher order which replays the principal features of the quantum-physical code, but is not reducible to it for fundamental reasons.

The reflexive symmetry of logic and quantum physics is in fact a fractal symmetry in which a solution to a wave equation serves as an input for logic operator equations:

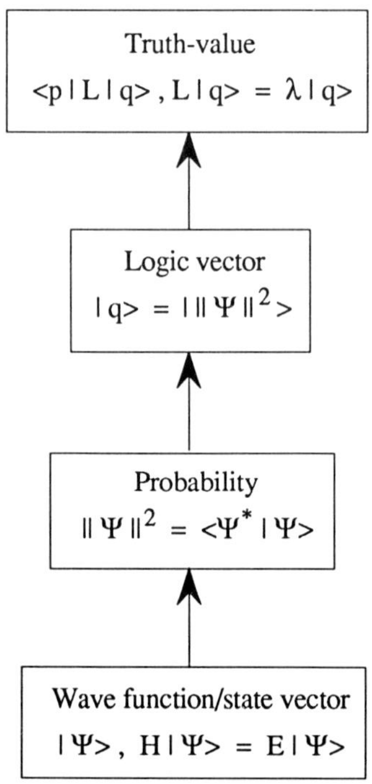

Fig.3 Fractal architecture of the intelligence code

The mind is a fractal of quantum mechanics, with quantum architecture fully recurring at the higher level of logic transformations.

This analysis, carried out to its logical conclusion, raises the fundamental question of whether cognition as higher-level quantum mechanics could have direct access to underlying covariant physical probabilities $\langle \Psi^* \mid \Psi \rangle$, thus being able to operate on a wave function directly. Neither the cognitive logic processes nor their codification and retrieval can be entirely random. Because high-level intelligence indeed has the capacity to control to a certain extent the process of logic deduction, could this imply that a wave function of the developed brain can be modified by the faculty of the mind? The answer to this question ultimately depends on whether cognition does or does not obey the linear laws of composition — the backbone both of classical and quantum physics.

The fact that a logic eigenstate of cognition cannot generally be obtained from another state through a linear transformation is a serious warning that linearity has to be given up in fractal quantum mechanics. The transitions in logic space violate the otherwise universal laws of linearity and consequently there is compelling reason to assume that the mind is indeed capable of altering the physical quantum wave packet probabilities of the brain. Consequently, the existing system of the covariant laws of physics must be extended to include fractal quantum mechanics.

The puzzle of the mind can be solved neither on the basis of classical notions nor on the basis of quantum notions. New fundamental concepts which go beyond the scope of both classical and quantum physics are required. Experimental confirmation of this result will have profound repercussions on both physical science and bioscience, as well as on science as a whole.

3.4 Annihilation and creation operators

3.4.1 Spacetime diagrams

According to quantum field theory all interactions arise from the exchange of particles. Each set of interactions can be structured in a spacetime map by vertices connected by oriented paths. The vertices are viewed as catastrophe points in spacetime where incoming particles are annihilated and new particles are created.

Since the time axis is directed from the bottom up, the paths directed backwards in time are associated with the antiparticles. The horizontal axis represents 3-D space. The diagrams below describe several abstract cases of interactions:

a)

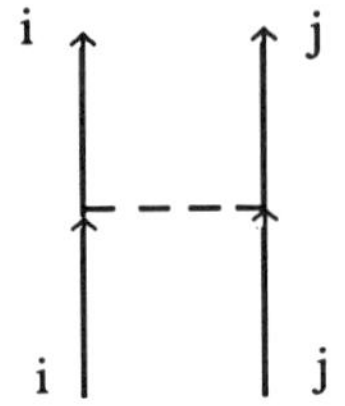

<ij | Q | ij>

b)

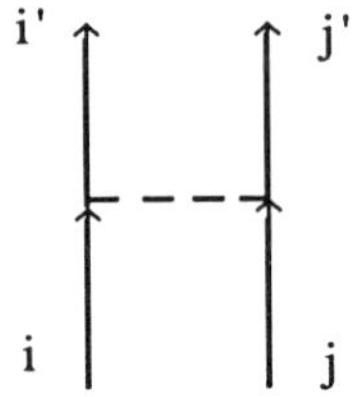

<i'j' | Q | ij>

c)

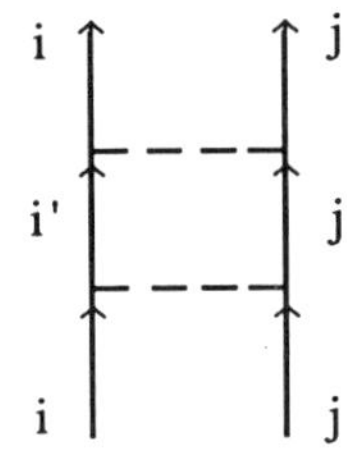

<ij | Q | i'j'><i'j' | Q | ij>

Diagram a) depicts two particles in the states i and j which as a result of interaction remain unchanged. Diagram b) depicts two interactions in which the initial states are dispersed into the new states i' and j'. Diagram c) describes two consecutive interactions. In the first interaction the initial states i and j are dispersed into the intermediary states i' and j' , and in the second interaction the initial states are recovered.

The concept of the annihilation and creation of particles is of fundamental significance for the spacetime diagrams which have an exact correspondence with the mathematical expressions of quantum field theory. The diagrams a) and b) with a single interaction are associated with the following matrix elements respectively:

<ij | Q | ij> and <i'j' | Q | ij>

where i, j, i', j' are eigenstate vectors, and Q is the quantum-mechanical operator.

Matrix elements associated with diagram c), which describes two consecutive acts of interaction, are written as a normal product:

<ij | Q | i'j'><i'j' | Q | ij> .

A parallel can be drawn between the normal products of quantum field theory and the fundamental product reductions we derived in matrix logic. In the same way as the complete diagram of fundamental interactions involves multiplication of normal products, a compound logic expression is also reducible to a fundamental product of the same multiplicative structure.

Compare for example the identity transformation in diagram c) with a decomposition of the equivalence product:

$$\langle ij \mid Q \mid i'j' \rangle \langle i'j' \mid Q \mid ij \rangle \qquad \text{and} \qquad \langle x \mid v \mid \bar{y} \rangle \langle \bar{x} \mid v \mid y \rangle$$

The multiplicative structure of both expressions is identical.

According to the reduction theorem of matrix logic any logic expression can be reduced to an algebraic sum of products of bracket products. Consequently, it might be possible, for any set of given diagrams, to establish a corresponding matrix logic expression and vice versa! When this has been effectively achieved, we can then describe fundamental interactions as logic networks, whereas on the other hand logic processes could be interpreted and studied with the help of physical methods.

So far logic has been developing without being grounded in fundamental principles. That approach, however, cannot reveal the relativistic covariance of the theory, which in itself is sufficient to demonstrate the absolute necessity of conceiving such a theory from initial principles. The affinity of matrix logic and fundamental interactions renders a synthesis of physics and logic into a unified theory entirely feasible.

The reformulation of physical theory in terms of fundamental logic promises to have a unifying effect of major significance.

Unitary symmetry

The construction of general symmetry group that can unify different interactions is an important objective of fundamental physics. The reason why the concept of symmetry is so powerful resides in the fact that the operations of symmetry are directly related to conservation laws. Whenever some property or quantity is conserved during an interaction, this is due to the underlying symmetry group. What is very important is that in modern theories the symmetry operations are considered not only in physical space but also in abstract mathematical spaces, applying both to a single particle or to a collection of particles. The introduction of logic rules of composition is proving to be an extremely useful tool in the search for higher-order symmetry. Considering a symmetry pattern for a multiplet of symmetric logic operators, we have established an octet frame which structurally mirrors the symmetry patterns of strong interactions:

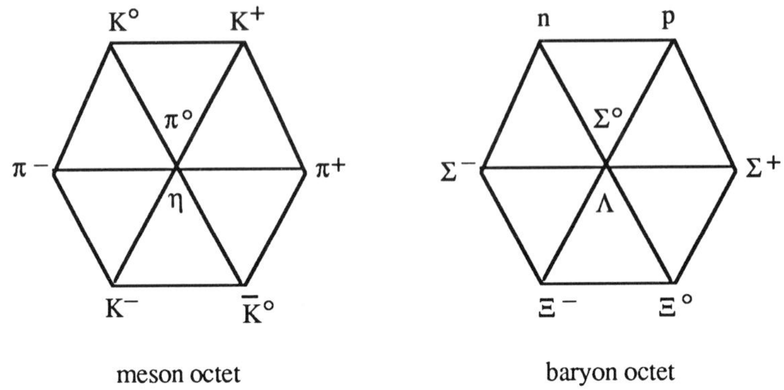

In the logic octet the diagonals connect pairs of complementary operators with

$$\mu(\bar{L}) = 4 - \mu(L) .$$

Similarly, in the meson octet particles and antiparticles occupy opposite positions in the octet, the two particles π° and η in the centre being their own antiparticles. For baryons the antiparticles are not contained in the octet, but form an identical anti-octet.
The symmetry pattern for the octet of non-symmetric logic operators, comprising a quadruplet of asymmetric and a quadruplet of bi-asymmetric operators,

$$\{\rightarrow, \leftarrow, \backslash, /, \perp, \mathrm{T}, \dashv, \vdash\} ,$$

takes the form of a square envelope pattern, with diagonals linking, as required, complementary operators, horizontals linking dual operators and verticals linking the pairs of transposes of the same logic norm,

$$\mu(L) = \mu(\tilde{L}) ,$$

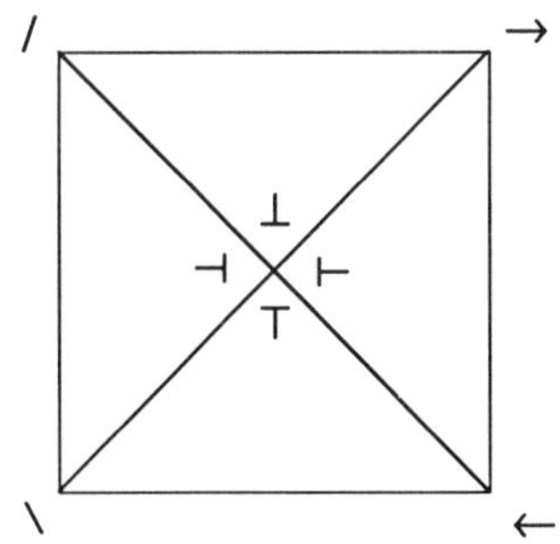

In the search for a global symmetry group that includes both octets of logic operators we are forced out of 3-D into 4-D space. The three-dimensional symmetry groups are insufficient to accommodate the complete set of logic operators. However, the complete logic multiplet, which remains invariant with respect to the triple of fundamental operations of symmetry,

$$L = \theta\,\tilde{\bar{L}}\ ,$$

is accommodated properly by the Zellweger hypercube, where the basic set of 16 logic operators is assigned to the precisely 16 vertices of the hypercube.

To grasp a hypercube logic architecture, assume that the vertices are labelled from 0 to $2^n - 1$ for some integer n. The integer n is the dimension of the hypercube, in our case, n = 4. The two logic operators are linked if and only if their binary codes differ in exactly one bit position. A 0-dimensional hypercube is just one vertex operator, and for n greater than 0 we define an n-dimensional hypercube as two (n–1)-dimensional hypercubes linked accordingly.
Each vertix in the hypercube communicates with the appropriate quadruplet of vertices, with transversing diagonals connecting pairs of complementary operators. Interpreting the global operations of cognition as 4-D rotations of the hypercube, we not only accommodate the higher level of abstraction but also, and more importantly, transform the vague psychological concept into a precise and experimentally verifiable theory. The fact that the total logic transformation is four-dimensional underscores indirectly the importance of the Lorentz invariancy for fundamental logic theory.
When intelligent operations are reduced strictly to the positive subset of logic, and if necessary negative and complex-valued operators can be handled in terms of several non-negative and real components, the logic system of the mind as a whole is clearly a hypercube, with the established invariancy rule

$$L = \theta \, \overset{\approx}{\bar{L}} \, ,$$

serving as a general law of conservation:

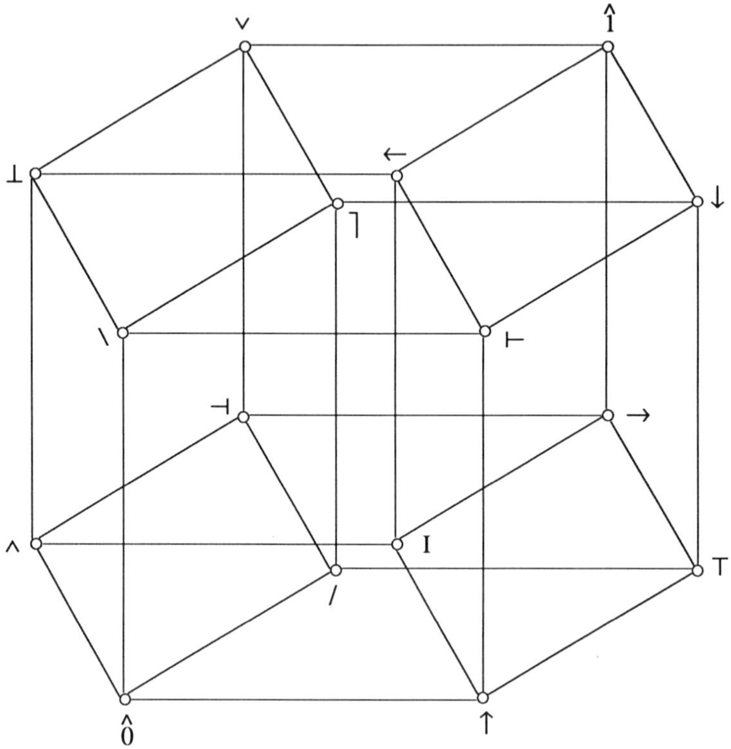

Fig. 4 Hypercube logic operator of cognition

Vertex complementation of the hypercube is equivalent to orthogonal rotation of the hypercube in 4-space with the change-over between constituent 3-cubes. This rotation replaces contradiction by tautology and negation by affirmation. We are in a position to employ the rotation of the hypercube for the purpose of logic evaluations.

Consider the problem of finding an adequate connective operator:

$$\langle x \mid ? \mid y \rangle .$$

The goal is to obtain a proper vertix operator to insert into the atomic product. Rotating the hypercube E^4 appropriately and then regressing it to a 0-dimensional hypercube E^0, we obtain a solution:

$$\circlearrowright$$
$$\langle x \mid E^4 \mid y \rangle$$
$$\Downarrow$$
$$\langle x \mid E^0 \mid y \rangle$$
$$\Downarrow$$
$$\langle x \mid L \mid y \rangle$$

Slicing off a pair of complementary vertices, we can detach with each vertex a quadruplet of operators, interconnected by the appropriate symmetry transform:

I $\{\wedge, \uparrow, \leftarrow, \rightarrow\}$	$\urcorner$ $\{/, \backslash, \downarrow, \vee\}$	$L_{\mathrm{I}} = \urcorner L_{\urcorner}$
$\wedge$ $\{\hat{0}, \mathrm{I}, \perp, \dashv\}$	$\downarrow$ $\{\hat{1}, \urcorner, \mathrm{T}, \vdash\}$	$L_{\wedge} = \bar{L}_{\downarrow}$
$\uparrow$ $\{\hat{0}, \mathrm{I}, \mathrm{T}, \vdash\}$	$\vee$ $\{\hat{1}, \urcorner, \perp, \dashv\}$	$L_{\uparrow} = \bar{L}_{\vee}$
$\backslash$ $\{\hat{0}, \urcorner, \perp, \vdash\}$	$\rightarrow$ $\{\hat{1}, \mathrm{I}, \mathrm{T}, \dashv\}$	$L_{\backslash} = \bar{L}_{\rightarrow}$
$/$ $\{\hat{0}, \urcorner, \mathrm{T}, \dashv\}$	$\leftarrow$ $\{\hat{1}, \mathrm{I}, \perp, \vdash\}$	$L_{/} = \bar{L}_{\leftarrow}$
$\perp$ $\{\wedge, \vee, \leftarrow, \backslash\}$	T $\{\uparrow, \downarrow, \rightarrow, /\}$	$L_{\perp} = \theta L_{\mathrm{T}}$
$\vdash$ $\{\uparrow, \backslash, \leftarrow, \downarrow\}$	$\dashv$ $\{\wedge, /, \rightarrow, \vee\}$	$L_{\vdash} = \theta L_{\dashv}$
$\hat{0}$ $\{\wedge, /, \backslash, \uparrow\}$	$\hat{1}$ $\{\vee, \rightarrow, \leftarrow, \downarrow\}$	$L_{\hat{0}} = L^{*}_{\hat{1}}$

Whenever a quadruplet displays a certain symmetry, there is a clearly defined logic operation which allows the logic property to be conserved with a transition from one quadruplet to another.

Application of the hypercube logic concept in fundamental physics offers the exciting prospect of resolving outstanding supersymmetry problems. Hypercube logic supersymmetry points vividly to the limitations of existing models and opens promising avenues towards the implementation of logic concepts in the theory of fundamental interactions.

The theory of fundamental forces which began with spin and isospin symmetry groups and ended with with the quark model can be extended in logic terms of composition, which provides a new alternative treatment of fundamental interactions.

Following the logic rules of composition, we can join the baryon octet and the baryon anti-octet into a single symmetry pattern. Similarly, 8 mesons and 8 baryons can be unified into a single hypercube. The fact that neither weak nor electromagnetic or gravitational forces have such a close correspondence with the logic hypercube highlights the importance of strong interactions for the fundamental theory of logic.

Virtual interactions

The subtraction or addition of the contradiction rule formulated in the previous chapter obtains fundamental meaning within the framework of the diagram method. Because the contribution of a contradiction in a diagram is averaged to zero, it can be equated to a virtual vacuum polarization:

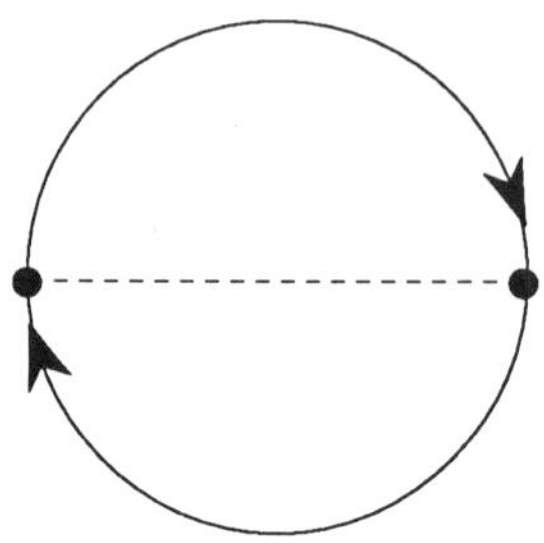

In virtual interactions the initial particles disintegrate into other particles only to re-emerge in accordance with the laws of conservation. In the vacuum transformation particles are created out of the void and vanish again into the void. When such a virtual loop is inserted into a logic string, we obtain a coupling corresponding to the fundamental rule of the subtraction or addition of a contradiction. In the section dealing with implication algebra we considered the logic equality

$$(x \to y) \to x = x .$$

The corresponding diagram contains the virtual interaction $\langle x | 1 \rangle \langle y | 1 \rangle \langle x | 0 \rangle$:

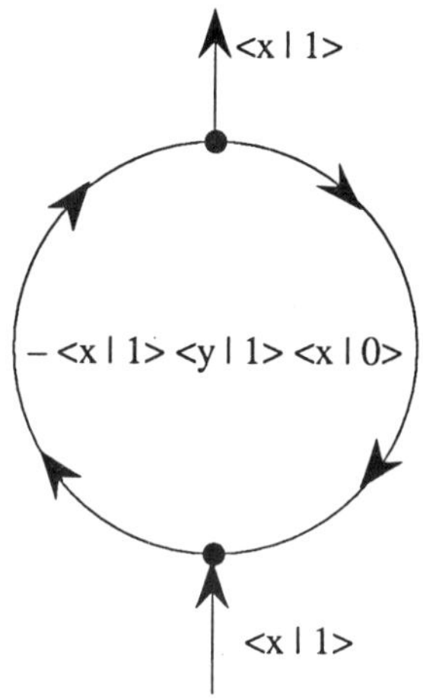

The logic state $|x\rangle$, which occurs at the front end of the interaction channel, must re-appear at the exit due to the conservation laws. The total virtual interaction is identity. However, if a virtual particle escapes the system, the output of the channel need not be $|x\rangle$. Within the limits imposed by the uncertainty principle a virtual logic state of the mind may tunnel into real awareness, providing a plausible explanation for the heuristic component of creativity.

3.4.2 Concept formation

Quantum theory offers an adequate description of single particles. For a many-particle system we have to resort to various approximation methods. In the first approximation the system of quantum particles is considered to comprise non-interacting particles. Then the wave function of the system Ψ is simply a product of the wave functions of individual particles:

$$\Psi = \Psi_1 \Psi_2 \Psi_3 \dots \Psi_n .$$

Similarly, in matrix logic, if logic states do not interact, a fundamental product is simply a product of bracket products:

$$\langle x_1 | L_1 | y_1 \rangle \langle x_2 | L_2 | y_2 \rangle \dots \langle x_n | L_n | y_n \rangle .$$

The introduction of interactions into a matrix logic string is accomplished by the complement operators, which bind certain clusters of bracket products together:

$$\langle \overline{x_1 | L_1 | y_1 \rangle \langle x_2 | L_2 | y_2} \rangle \dots \langle x_n | L_n | y_n \rangle .$$

In psychological terms, this can be viewed as the formation of associations or concept formation. In mathematical terms, this leads to the formation of an algebraic sum of products, where individual bracket products are allowed to float in a cluster under a complement but cannot be separated from one another. What is described in logic categories as concept formation can serve as a mathematical model for coupling wave functions in a many-particle interaction system and vice versa.

Concept formation in a quantum-physical framework is equivalent to the vectorization of particles. The bracket products under the complement operator (scalars) are translated into the verum components of the newly formed logic vectors:

$$<\overline{x_1 \mid L_1 \mid y_1 >< x_2 \mid L_2 \mid y_2}> \; = \; << x_1 \mid L_1 \mid y_1 > \mid \downarrow \mid < x_2 \mid L_2 \mid y_2 >> \; ,$$

where

$$<< x_1 \mid L_1 \mid y_1 > \mid \; = \; \left(<\overline{x_1 \mid L_1 \mid y_1}> \; , \; < x_1 \mid L_1 \mid y_1 > \right) ,$$

$$\mid < x_2 \mid L_2 \mid y_2 >> \; = \; \begin{pmatrix} <\overline{x_2 \mid L_2 \mid y_2}> \\ < x_2 \mid L_2 \mid y_2 > \end{pmatrix} .$$

Whenever vectorization of particles is possible, the translation of a product of bracket products into a single atomic product can be carried out. The trade-off for this metamorphosis is an increase in the local internal organization of the system.

3.4.3 Bosons and fermions

The fundamental postulate of quantum theory states that the operators of coordinate q and momentum p satisfy the commutation relation

$$[q, p] \; = \; qp - pq \; = \; i\hbar$$

The effect of the application of a commutator to a wave Ψ function equals the same wave function multiplied by a constant $\hbar$:

$$[q, p]\,\Psi \; = \; i\hbar\Psi$$

To obtain the quantized energy levels in agreement with the commutation postulate, the energy operator

$$H = \frac{1}{2}(p^2 + \omega^2 q^2)$$

is rephrased in terms of two special operators, the creation operator a* and the annihilation operator a:

$$a = \frac{1}{\sqrt{2\hbar\omega}}(\omega q + ip)$$

$$a^* = \frac{1}{\sqrt{2\hbar\omega}}(\omega q - ip)$$

where a and a^* are complex conjugates, ω is frequency, $\hbar$ is the action quantum. The operators q and p are then expressed as follows:

$$q = \sqrt{\frac{\hbar}{2\omega}}\left(a^* + a\right)$$

$$p = i\sqrt{\frac{\hbar\omega}{2}}\left(a^* - a\right)$$

As with q and p, the operators a and a^* do not commute. Substituting them into the commutator,

$$[q, p] = i\hbar \ ,$$

we obtain the fundamental commutation relation for annihilation and creation operators

$$[a, a^*] = 1 \ .$$

In terms of annihilation and creation operators, the energy operator takes the following form:

$$H = \frac{\hbar\omega}{2}(aa^* + a^*a) = \hbar\omega(a^*a + \tfrac{1}{2})$$

The annihilation operator a^* decreases the number of quanta in a system, while the operator a creates energy quanta. The product a^*a is an operator which represents the number of quanta in a given state:

$$N = a^*a \ .$$

From the fundamental relation

$$[a, a^*] = 1$$

we derive

$$[a, a^*a] = a \,,$$
$$[a^*, a^*a\,] = -a^*$$

and, employing operator N, we get

$$Na = a\,(N-1)$$
$$Na^* = a^*\,(N+1)\;.$$

Considering the orthonormal and complete system of states

$$\langle n \mid n' \rangle = \delta_{nn'}$$

$$\sum_{n=0}^{\infty} \mid n \rangle \langle n \mid \;=\; I$$

we can define them by the relations

$$a \mid n \rangle \;=\; \sqrt{n} \mid n-1 \rangle$$

$$a^* \mid n \rangle \;=\; \sqrt{n+1} \mid n+1 \rangle \;.$$

The state $\mid n\rangle$ can be interpreted as a state comprising n quanta; then $\mid 0\rangle$ is a vacuum state. Application of the creation operator a^* to a state $\mid n\rangle$ with n quanta produces a state $\mid n + 1\rangle$ with n + 1 quanta. Similarly, application of the annihilation operator a to a state $\mid n\rangle$ produces a state $\mid n - 1\rangle$ with n – 1 quanta.
The quantum numbers n are all positive integers or 0:

$$n = 0,\, 1,\, 2,\, \ldots$$

This implies that the values of energy are quantized:

$$E_n = \hbar\omega\,(n + \tfrac{1}{2})$$

For large values of n the quantized character of energy is suppressed and the quantum relation converges to classical laws.
Observation of the operator N gives one of its eigenvalues

$$0,\ 1,\ 2,\ \ldots,\ \infty,$$

which characterizes the number of quanta or particles in a given state. According to such an interpretation, there can in any state be an arbitrary number of particles. The particles characterized by this property are bosons, satisfying the fundamental commutation relation

$$[a, a^*] = 1\ ,$$

The boson commutator $[a, a^*]$ is equal to the unit operator; its application to a state reproduces the state:

$$\begin{aligned}
[a, a^*] \mid n\rangle &= \\
&= a\, a^* \mid n\rangle - a^*\, a \mid n\rangle = \\
&= a \sqrt{n+1} \mid n+1\rangle - a^* \sqrt{n} \mid n-1\rangle = \\
&= \sqrt{n+1}\, a \mid n+1\rangle - \sqrt{n}\, a^* \mid n-1\rangle = \\
&= \sqrt{n+1} \sqrt{n+1} \mid n\rangle - \sqrt{n} \sqrt{n} \mid n\rangle = \\
&= (n+1) \mid n\rangle - n \mid n\rangle \\
&= \mid n\rangle\ .
\end{aligned}$$

It is important for further discussion that a state can be obtained by application of the creation operator to a vacuum state $\mid 0\rangle$:

$$a^* \mid 0\rangle = \mid 1\rangle\ ,$$

whereas application of the annihilation operator to a vacuum state gives 0:

$$a \mid 0\rangle = 0\ .$$

This implies that the quantum numbers n cannot take negative values!

Boson logic states

The fundamental commutator relating boson annihilation and creation operators,

$$[a, a^*] = 1 ,$$

has a corresponding counterpart in logic. While the boson commutator is a consequence of the fundamental postulate

$$[x, p] = i\hbar \quad ,$$

for logic bosons the commutator is not postulated but is derived from general logic theory.
To begin with, we identify the empty vacuum logic state with a false state 0. Then, in order to create a state from the vacuum logic state, we can employ the complement operator to serve as a creation operator:

$$\bar{0} = 1$$

We shall now seek an adjoint operator which will allow us to annihilate the true state 1 or the verum state into the empty state 0, and also when operating on the false vacuum state to yield the false state again. In attributing such properties to the annihilation operator, we shall be faithful to the boson annihilation operator in physics.
The operator satisfying our requirement is the differentiation operator $'$. Indeed,

$$\bar{0} = 1 \qquad — \qquad \text{creation}$$
$$1'_p = 0 \qquad — \qquad \text{annihilation}$$

If the logic state is empty, application of the annihilation operator $'$ cannot do anything.
The use of the differential operator $'$ as the annihilation operator is justified because boson logic states are characterized by a continuous spectrum. In logic interactions the adjoint to the annihilation operator is not, however, the integration operator, but the complement operator.
With the logic operators of annihilation and creation defined above, we may now construct a tautology commutator for the boson logic states:

$$[^-,'] \, L(p, q) = \overline{L'(p, q)} - \overline{L(p, q)}\,'$$

The commutator accommodates the reflexive symmetry of logic and fundamental interactions: the operators of creation and annihilation in the logic commutator stand in the reverse order to that of quantum-physical description. Application of the commutator to any logic function is invariably a tautology, regardless of the variable of differentiation:

$$[^-,'] \, L(p, q) = 1 ,$$

and

$$\left[-,\frac{d}{dp}\right]L(p,\,q)=\left[-,\frac{d}{dq}\right]L(p,\,q)$$

Fermion logic states

According to the Pauli exclusion principle, the fundamental property of fermions is that a given quantum state can be occupied only by one fermion. No two fermions can ever occupy the same quantum state. Because of this restriction the fundamental commutation relation

$$[a,\,a^*]=aa^*-a^*a=1$$

is inapplicable to fermion states. However, the theory of fermions can be designed in a way similar to that of bosons, if the commutation relation for the creation and annihilation operators is replaced by the Wigner anticommutator:

$$\{a,\,a^*\}=aa^*+a^*a=1\ ,$$

where a and a^* are now fermion annihilation and creation operators respectively.

In contrast with the operators a and a^* for bosons, the fermion operators of annihilation and creation are not connected with the operators of coordinate and momentum. In addition, compliance with the following relations is required:

$$\{a,\,a\}=0$$

$$\{a^*,\,a^*\}=0$$

$$a^2=a^{*2}=0\ .$$

The operator

$$N=a^*a\ ,$$

characterizing the number of fermions in a given state, will adequately describe a fermion state if it has only the two eigenvalues 0 and 1, so that a state is either empty or occupied by a single fermion.

For fermions, characterized by a discrete spectrum, we shall use, instead of the continuous logic functions L(p,q), discrete logic vectors and discrete operators L, i.e. we are in a discrete universe.

Whereas in the case of logic bosons we employed the complement operator as a creation operator but had to use differentiation as an annihilation operator, now both creation and annihilation operators can, in view of the complementarity of fermion logic states, be modelled on the complement operator.
For the creation of the true state we have

$$|\bar{0}> = |1>$$

and for its annihilation

$$|\bar{1}> = |0> .$$

The complement acts here both as a creation operator and as an annihilation operator.
In order to replace the complement operator by equivalent multiplication operators that would have the same effect, we employ implication $\rightarrow$ and converse implication $\leftarrow$. These act upon the false state $|0>$ and the true state $|1>$ respectively as invariant identity operators:

$$\rightarrow |0> = |0>$$

$$\leftarrow |1> = |1>$$

This is to be expected since the logic eigenvalue for both $\rightarrow$ and $\leftarrow$ is 1.
After establishing these equivalences, we shall now return to the initial equations for creation and annihilation:

$$|\bar{0}> = |1>$$

$$|\bar{1}> = |0> ,$$

and rewrite them so as to include the invariant implication operations. We have

$$|\bar{0}> = \overline{\rightarrow |0} > = |1> ,$$

$$|\bar{1}> = \overline{\leftarrow |1} > = |0> .$$

Employing the familiar rule of matrix logic, we eliminate the complement operators:

$$\overline{\rightarrow |0} > = \backslash |0> = |1>$$

$$\overline{\leftarrow |1} > = / |1> = |0> ,$$

and obtain a multiplicative matrix operator form for the creation \ and annihilation / operators:

$$\backslash = \begin{pmatrix} 0 & 0 \\ 1 & 0 \end{pmatrix} \quad \text{— } a^* \text{– operator}$$

$$/ = \begin{pmatrix} 0 & 1 \\ 0 & 0 \end{pmatrix} \quad \text{— } a \text{– operator}$$

Consequently, we can present the operator N characterizing the number of fermions in a quantum state as a logic operator of conjunction:

$$N = a^*a = \backslash\, / = \wedge\,.$$

The two eigenvectors of $\wedge$ corresponding to the two eigenvalues 0 and 1 are the false and true logic vectors:

$$|\,0> = \begin{pmatrix} 1 \\ 0 \end{pmatrix} \quad \text{and} \quad |\,1> = \begin{pmatrix} 0 \\ 1 \end{pmatrix}.$$

The eigenvectors satisfy the orthonormality and closure relations:

$$<0\,|\,0> = <1\,|\,1> = 1$$

$$<0\,|\,1> = <1\,|\,0> = 0$$

$$|\,0><0\,| + |\,1><1\,| = I$$

All this is quite similar to the case of bosons, with the major difference, however, that in the eigenequation

$$\wedge\,|\,n> \;=\; n\,|\,n>$$

the number n is allowed to have only two values, 0 and 1.

The creation operator \ applied to state $|\,0>$ will produce the state $|\,1>$:

$$\backslash\,|\,0> \;=\; |\,1>$$

Because

$$\backslash^2 = \hat{0}$$

we automatically exclude the appearance of two fermions in the same state.
Similarly, the annihilation operator applied to the occupied fermion state produces an empty state,

$$/ \,|\, 1\rangle = |\, 0\rangle$$

and again

$$/^2 = \hat{0}\,.$$

The converse inhibition operator / describes annihilation adequately. Whenever the inhibition operator is applied to an occupied state or converse inhibition to an empty state, in both cases

$$\backslash \,|\, 1\rangle = |\, \mathbf{0}\rangle$$

$$/ \,|\, 0\rangle = |\, \mathbf{0}\rangle\ .$$

This is to be expected. According to the exclusion principle, a fermion cannot be created if a state is already occupied; while it is evident that a particle cannot be annihilated if it is absent. Thus for fermion logic, as for boson logic, we have

$$/ \,|\, n\rangle = \sqrt{n}\,|\, n-1\rangle$$

$$\backslash \,|\, n\rangle = \sqrt{1-n}\,|\, n+1\ ,$$

but here the quantum numbers n vanish for values greater than 1 and smaller than 0.

The fact that the quantum algebra of annihilation and creation operators is adequately accommodated by logic operators is an indication of a reflexive symmetry between the nature of logic and the logic of nature. The fusion of matrix logic methods with the methods of fundamental physics changes the way we interpret the phenomena of the mind, making it possible to introduce logic descriptions into physical theory and physical description into logic science. In this manner the matrix operator formulation of logic both supports an important breakthrough in the study of the physics of the mind and exemplifies the way logic theory can be brought to bear on the theory of fundamental interactions.

3.4.4 A logic model of matter

The introduction of logic operators into physical description and of physical operators into logic description opens an avenue for the development of a logic model of matter, in which the structural conservation laws of physics and of logic become two aspects of reality with reflexive fractal symmetry. In physics symmetry has established itself as one of the basic concepts. Though symmetry relations are deeply present at the very foundation of logic, they have not been considered as occupying a focal point there. Matrix logic places this problem in a new perspective, where logic operations can be formalized as operations of symmetry in logic space.

The fractal architecture of the symmetry groups in logic and physics, in spite of the different nature of the group elements, allows us to study these processes either from a physical viewpoint or from the point of view of logic.

Spin

Spin is a fundamental quantum quantity which is not derivable from classical notions. A complete definition of a state of a particle requires not only a specification of its space and time coordinates but also of the direction of the spin vector. Consequently, the wave function of a particle becomes dependent on the spin variable which can be assigned to it as an index, specifying the direction of spin, either up or down.

The arbitrary spin state is represented by a spinor or two-component wave function, quantized into spin-up $\psi_\uparrow$ and spin-down $\psi_\downarrow$ components:

$$\psi = \begin{pmatrix} \psi_\uparrow \\ \psi_\downarrow \end{pmatrix}.$$

Squares of modulo $|\psi_\uparrow|^2$ and $|\psi_\downarrow|^2$ give the probabilities of observing transitions into the states $\psi_\uparrow$ and $\psi_\downarrow$ respectively, and the sum of the probabilities is unity:

$$|\psi_\uparrow|^2 + |\psi_\downarrow|^2 = 1.$$

Spin operators have two eigenstates and are expressed by the set of Pauli matrices, which together with the identity matrix form the unitary group of symmetry:

$$\sigma_0 = \begin{pmatrix} 1 & 0 \\ 0 & 1 \end{pmatrix} \qquad \sigma_x = \begin{pmatrix} 0 & 1 \\ 1 & 0 \end{pmatrix} \qquad \sigma_y = i\begin{pmatrix} 0 & -1 \\ 1 & 0 \end{pmatrix} \qquad \sigma_z = \begin{pmatrix} 1 & 0 \\ 0 & -1 \end{pmatrix},$$

with cyclic closure

$$\sigma_x\sigma_y = -i\sigma_z\,,\ \ \sigma_y\sigma_z = -i\sigma_x\,,\ \ \sigma_z\sigma_x = -i\sigma_y\,,$$

anticommutation

$$\sigma_x\sigma_y + \sigma_y\sigma_x = 0$$

$$\sigma_y\sigma_z + \sigma_z\sigma_y = 0$$

$$\sigma_z\sigma_x + \sigma_x\sigma_z = 0\,,$$

and

$$\sigma_x^2 = \sigma_y^2 = \sigma_z^2 = \sigma_0\,.$$

Spinor transformations occur in unitary spinor space with the spinor having the most simple forms when the quantizing axes coincide:

$$\psi_\uparrow = \begin{pmatrix} 1 \\ 0 \end{pmatrix} \quad \text{— for spin } \tfrac{1}{2}$$

$$\psi_\downarrow = \begin{pmatrix} 0 \\ 1 \end{pmatrix} \quad \text{— for spin } -\tfrac{1}{2}\,.$$

These spinors are the basis vectors which correspond to the basis spinor states.

The affinity of matrix logic and spin is striking. The logic operators, just like spin operators, can have no more than two eigenvalues. The basis states of the spinors are precisely the false and true logic vectors

$$\Psi_\uparrow = |0\rangle \quad \text{and} \quad \Psi_\downarrow = |1\rangle$$

which we employed in a discrete universe. As a matter of fact, the two orthogonal quantum states of a spinor can be uniquely mapped onto the macroscopically orthogonal true and false logic states. It is only in this case that the normalization condition for a spinor coincides with the condition for normalization of logic vectors. A pure state of a spinor is identical with a base logic vector.

In the view that spin is the quantity which is not derivable from the classical notions and for which the macroanalogue remains unknown the possibility of matrix logic serving as such a macroanalogue cannot be excluded. In such an event, it might be possible not only to draw parallels between the logic processes and the spinor transformations, but to relate both in a fundamental sense.

The proposal to view discrete logic as a macroanalogue of spin is reinforced by the essential properties of the negation operator. Basically, mental division is twofold. The mind tends to formulate the results of logic evaluation in the unambiguous and mutually exclusive terms of true and false. We encounter great difficulties whenever it is necessary to accept or to communicate any data or information that are not definitely true or false. This gives rise to the hypothesis that spin provides a foundation for the macroscopically observed property of negation.

The two most fundamental logic operators, negation $\urcorner$ and tautology $\hat{1}$, clearly display spinor properties. The eigenstates of negation and tautology,

$$|q_1> = \begin{pmatrix} \frac{1}{2} \\ \frac{1}{2} \end{pmatrix} \qquad |q_2> = \begin{pmatrix} \frac{1}{2} \\ -\frac{1}{2} \end{pmatrix},$$

can be put in direct correspondence with the spin formalism. Moreover, the negation operator is in itself the σ_x Pauli matrix employed in the spin formalism.

Diagonalization of negation with the help of its eigenvectors

$$\begin{pmatrix} \frac{1}{2} \\ \frac{1}{2} \end{pmatrix} \quad \text{and} \quad \begin{pmatrix} \frac{1}{2} \\ -\frac{1}{2} \end{pmatrix}$$

gives us the σ_z component of the spin operator

$$\sigma_z = \begin{pmatrix} 1 & 0 \\ 0 & -1 \end{pmatrix} = \begin{pmatrix} 1 & 1 \\ 1 & -1 \end{pmatrix} \begin{pmatrix} 0 & 1 \\ 1 & 0 \end{pmatrix} \begin{pmatrix} \frac{1}{2} & \frac{1}{2} \\ \frac{1}{2} & -\frac{1}{2} \end{pmatrix},$$

The matrix $\sigma_z = \begin{pmatrix} 1 & 0 \\ 0 & -1 \end{pmatrix}$ has a diagonal form, which is a general property of any operator defined in the basis formed from its eigenvectors.

The spinor $|1>$ corresponding to true and the spinor $|0>$ corresponding to false form an orthogonal set:

$$<1|0> = 0 \qquad \text{and} \qquad <0|1> = 0 .$$

In order to establish a link between logic transformations and spinor transformations we employ a commutator, either

$$\frac{1}{2}\blacktriangledown = \frac{1}{2}[\wedge,\vee]$$

or

$$\frac{1}{2}\blacktriangle = \frac{1}{2}[\vee,\wedge] .$$

The corresponding bracket product in the discrete universe then yields two projections of logic spin

$$\frac{1}{2}<x|[\wedge,\vee]|y> = \begin{cases} \frac{1}{2} & \text{for } <x|\bar{y}> \\ -\frac{1}{2} & \text{for } <\bar{x}|y> \end{cases}$$

depending on the truth status of the orthogonal vectors $|x>$ and $|y>$.
Consequently, logic spin can perform an important operation of comparison. Recombining the operators $\wedge$ and $\vee$, we record an inversion of spin orientation:

$$\frac{1}{2}<x|[\wedge,\vee]|y> = \frac{1}{2}<x|-[\wedge^*,\vee^*]|y> = -\frac{1}{2}<x|[\vee,\wedge]|y>$$

This inversion is achieved through a transition to dual operators:

$$\wedge^* = \vee$$
$$\vee^* = \wedge$$

Because for dual logic operators

$$[\bar{L},\bar{L}^*] = -[L,L^*] ,$$

we can also employ, instead of the duality transformation, the complement operator. Complementation of operators changes the sign of the commutator to the opposite, as does the duality transformation.

Unitary transformations in spinor space are relativistically invariant. In the relativistic case we replace the 2-component Pauli spinor with the 4-component Dirac spinor:

$$\psi = \begin{pmatrix} \psi_1 \\ \psi_2 \\ \psi_3 \\ \psi_4 \end{pmatrix} ; \qquad \psi^* = \begin{pmatrix} \psi_1^* & \psi_2^* & \psi_3^* & \psi_4^* \end{pmatrix}$$

and the Pauli matrix base $\sigma_0, \sigma_x, \sigma_y, \sigma_z$ by Dirac (4×4) spinor matrices:

$$\gamma_1 = \begin{pmatrix} 0 & \sigma_x \\ -\sigma_x & 0 \end{pmatrix} \quad \gamma_2 = \begin{pmatrix} 0 & \sigma_y \\ -\sigma_y & 0 \end{pmatrix} \quad \gamma_3 = \begin{pmatrix} 0 & \sigma_z \\ -\sigma_z & 0 \end{pmatrix} \quad \gamma_4 = \begin{pmatrix} \sigma_0 & 0 \\ 0 & -\sigma_0 \end{pmatrix} .$$

Altogether the Dirac matrices $\gamma_1, \gamma_2, \gamma_3, \gamma_4$ generate a matrix base of 16 linearly independent quantities in which any (4×4) matrix can be expanded:

$$\begin{aligned}
&\gamma_0 = I , && \gamma_{10} = i\gamma_2\gamma_4 , \\
&\gamma_1, \gamma_2, \gamma_3, \gamma_4, \gamma_5 = \gamma_1\gamma_2\gamma_3\gamma_4 , && \gamma_{11} = i\gamma_3\gamma_4 , \\
&\gamma_6 = i\gamma_1\gamma_2 , && \gamma_{12} = i\gamma_1\gamma_2\gamma_3 , \\
&\gamma_7 = i\gamma_2\gamma_3 , && \gamma_{13} = i\gamma_2\gamma_3\gamma_4 , \\
&\gamma_8 = i\gamma_3\gamma_1 , && \gamma_{14} = i\gamma_3\gamma_1\gamma_4 , \\
&\gamma_9 = i\gamma_1\gamma_4 , && \gamma_{15} = i\gamma_1\gamma_2\gamma_4 .
\end{aligned}$$

Under the interpretations given, the duality operations in the logic vector space correspond to unitary transformations in the spinor space, which in turn are associated with corresponding Lorentz transformations in spacetime:

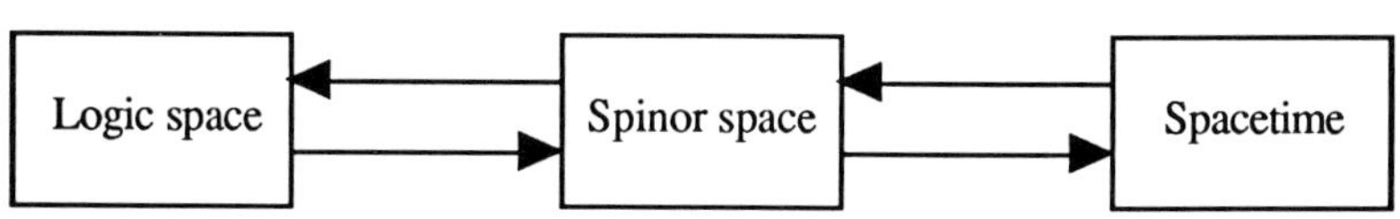

The duality transformations, unitary spinor transformations and the Lorentz group of linear transformations, which have found fundamental applications in relativity and quantum physics, can all be conveniently handled by the (2×2) matrix group, making the link with matrix logic quite simple and meaningful. The common mathematical roots of these three cornerstones of scientific inquiry are an indication that conservation laws in physics and in logic are not just two aspects of reality with reflexive symmetry between them, but are actually united in a common relativistically covariant framework.

In the previous chapter we showed that the problem of the functional completeness of logic can be addressed in algebraic terms as the problem of the existence of a complete matrix base. The proposal to link logic and spin derives additional support from the fact that the elements σ_t, σ_x, σ_y, σ_z of the spin group constitute a linear independent matrix base from which any (2×2) matrix can be generated, and conversely, given an operator, we can uniquely decompose it into the elements of the spinor matrix base with corresponding multipliers. In the relativistic case the matrix base is generated by the Dirac (4×4) spinor matrices. In the context of spinor logic the concept of the logic base receives fundamental elaboration. The fundamental logic properties are not derivable from classical notions, just as the properties of spin cannot be explained by classical theory.

Periodic table

The Pauli exclusion postulate accounts elegantly for the distribution of electrons in the orbits of atoms, which defines the place of the element in the periodic table. According to the exclusion postulate no two electrons with parallel spins can occupy the same quantum state. To allow the coupling of electrons in a given orbit, the postulate requires that they possess antiparallel spins.

A plausible interpretation of this phenomenon can be given from the matrix logic point of view. In logic space the addition of logic vectors must not violate the binarity conditions. Thus addition is not possible unless:

$$\bar{q}_1 + \bar{q}_2 \leq 1$$

$$q_1 + q_2 \leq 1$$

This effectively prohibits couplings such as

$$\begin{pmatrix} 0 \\ 1 \end{pmatrix} + \begin{pmatrix} 0 \\ 1 \end{pmatrix} = \begin{pmatrix} 0 \\ 2 \end{pmatrix}$$

or

$$\begin{pmatrix}1\\0\end{pmatrix} + \begin{pmatrix}1\\0\end{pmatrix} = \begin{pmatrix}2\\0\end{pmatrix},$$

because the resulting vectors are logically inadmissible. However, the coupling of orthogonal logic vectors is accepted without constraint:

$$\begin{pmatrix}0\\1\end{pmatrix} + \begin{pmatrix}1\\0\end{pmatrix} = \begin{pmatrix}1\\1\end{pmatrix}.$$

If we associate a given spin orientation of an electron with the true logic vector | 1> and the antiparallel spin orientation of another electron with the false logic vector | 0>, then an antiparallel electron pair in orbit is the logic vector sum of both vectors.
This concept is visualized below for a selected set of elements from the periodic table, where the increase in the number of electrons is controlled by the law of the addition of logic vectors which prevents the appearance of forbidden values.

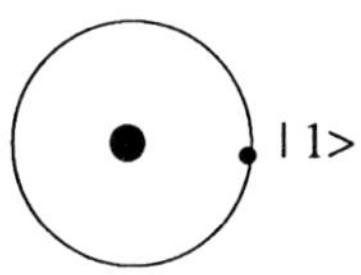

— H

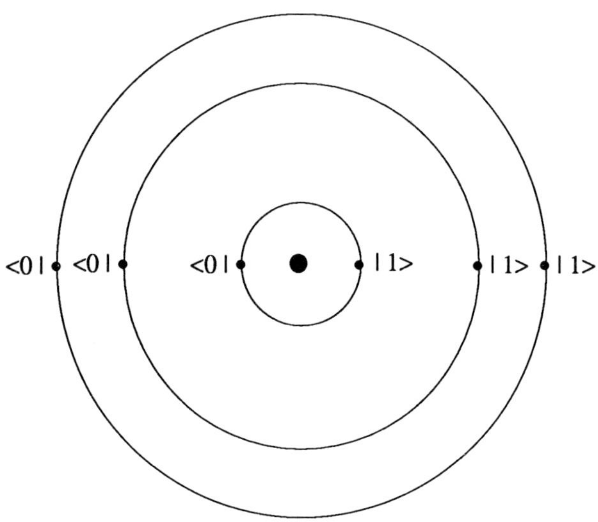

— C

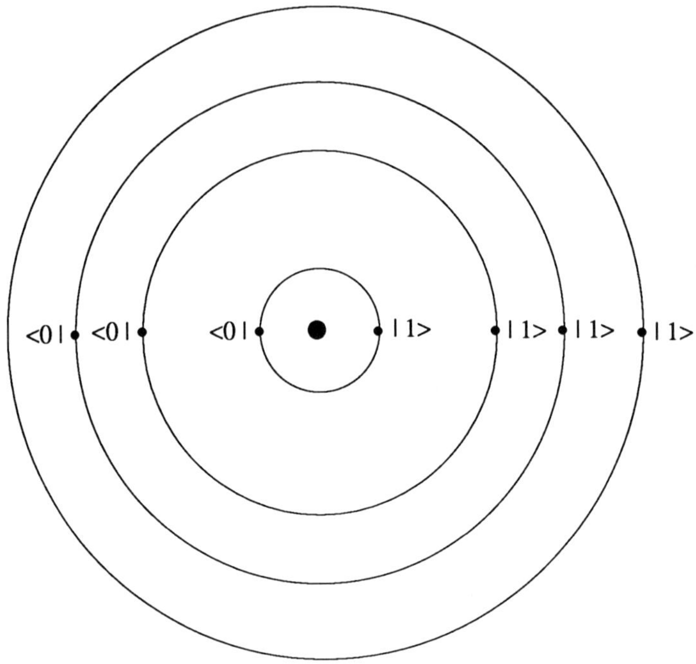

— N

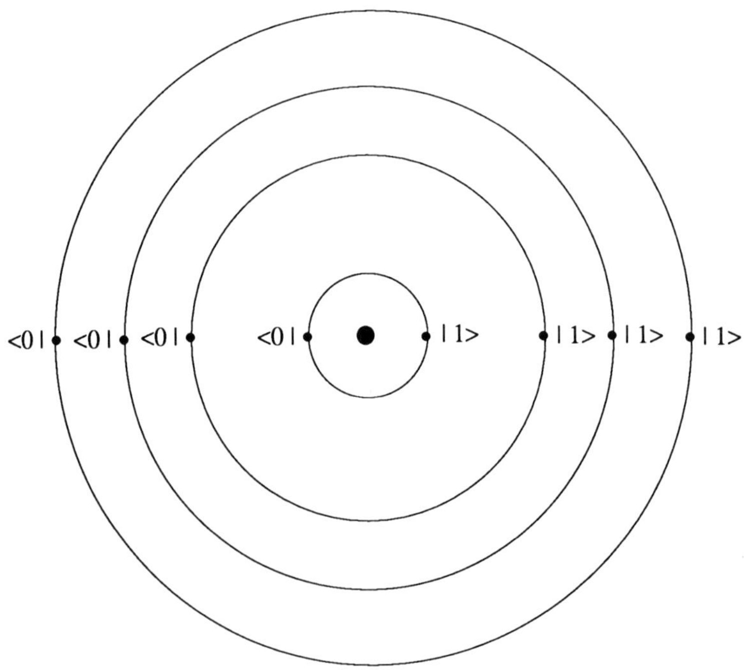

—O

The logic model of electron orbits leads to an interesting conclusion: if the Pauli exclusion postulate is interpreted as a consequence of the law of the addition of logic vectors, then what is postulated in quantum mechanics can be seen as a consequence of the natural requirement of logic theory.

The acceptance of the proposal that the fundamental structure of matter is controlled by the same laws as logic processes can contribute fruitfully to ongoing research into fundamental interactions, and especially to the classification of quantum fields and particles.

Isospin

From the logic point of view an atom can be interpreted as a codeword, comprising the electron codeword and the corresponding nucleon codeword. Of particular importance are isotopically invariant atoms, such as C, N, O, P, which together with hydrogen atoms H form the bulk of the brain matter:

shells	↑	↓	↑	↓	↑	↓	↑	↓	electrons
nucleus	+	0	+	0	+	0	+	0	nucleons
	0	+	0	+	0	+	0	+	

Oxygen codewords

The codewords for isotopically invariant atoms contain no empty cells, satisfying the important requirement of chemical stability and casting light on the fundamental puzzle how the organized system of atoms, in the end quarks, is able to think. Considering separately the nucleon codeword, we can distinguish two logic states of a nucleon, charged proton and neutral neutron:

$$\mathrm{N} := \begin{cases} 1 & - \text{ proton p} \\ 0 & - \text{ neutron n} \end{cases}$$

In matrix logic terms the two states of the nucleon are described by the basis true and false logic vectors:

$$\mathrm{N}^{+} := |\, 1 > \; = \begin{pmatrix} 0 \\ 1 \end{pmatrix}$$

and

$$N^0 := |\,0> \; = \begin{pmatrix} 1 \\ 0 \end{pmatrix}.$$

In isospin theory a proton and a neutron are considered to be two states of the same particle, the nucleon. The operators of the theory resemble spin operators. Here the number of states is also equal to two, but the physical quantity associated with these states is not spin but a charge Q. The charge obtains integer values, 0 for neutron, and 1 for proton. Isospin operators are represented by the same matrices as the spin operators

$$C_0 = \frac{1}{2}\begin{pmatrix} 1 & 0 \\ 0 & 1 \end{pmatrix} \quad C_1 = \frac{1}{2}\begin{pmatrix} 0 & 1 \\ 1 & 0 \end{pmatrix} \quad C_2 = \frac{i}{2}\begin{pmatrix} 0 & -1 \\ 1 & 0 \end{pmatrix} \quad C_3 = \frac{1}{2}\begin{pmatrix} 1 & 0 \\ 0 & -1 \end{pmatrix}$$

The charge operator Q must be introduced in such a manner that its eigenvalues are equal to 0 and 1:

$$Q = C_0 - C_3 = \frac{1}{2}\begin{pmatrix} 1 & 0 \\ 0 & 1 \end{pmatrix} - \frac{1}{2}\begin{pmatrix} 1 & 0 \\ 0 & -1 \end{pmatrix} = \begin{pmatrix} 0 & 0 \\ 0 & 1 \end{pmatrix}$$

All isospin operators are defined in the orthonormal basis. In the same basis the eigenvectors of the operator Q are

$$(1, 0) \text{ and } (0, 1)\,.$$

Thus the eigenvectors and the basis vector are the same:

$$Q\,|\,0> \; = 0\,|\,0>, \qquad Q\,|\,1> \; = 1\,|\,1>.$$

The possible values of the charge are the eigenvalues 0 and 1 of the operator Q.
The operators

$$C_+ = C_1 + i\,C_2 = \begin{pmatrix} 0 & 1 \\ 0 & 0 \end{pmatrix} = /$$

$$C_- = C_1 - i\,C_2 = \begin{pmatrix} 0 & 0 \\ 1 & 0 \end{pmatrix} = \backslash$$

translate the vectors of the proton and neutron states into each other:

$$C_+ | 1> \ = \ | 0>$$

$$C_- | 0> \ = \ | 1>$$

The change-over from proton to neutron and vice versa, occurring in the strong interactions, allows for interpretation of these physical processes in terms of logic switch operations. In the strong switch scheme presented below, the arrows indicate absorption and emission routes.

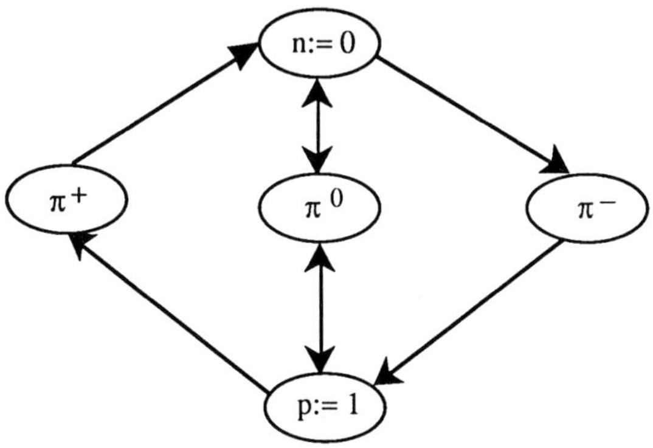

Ternary strong switch

The change-over proton-neutron is provided by interactions with the meson triplet $\{\pi^+ \ \pi^0 \ \pi\}$. A proton accepting π^- or emitting π^+ is translated into a neutron; a neutron accepting π^+ or emitting π^- is translated into a proton. Interactions with π^+ and π^- represent oriented negations, and the equations of composition and decomposition will thus be:

n:

$$(1,0)\begin{pmatrix} 0 & 1 \\ 0 & 0 \end{pmatrix} = (0,1) \qquad \text{– composition equation}$$

$$\begin{pmatrix} 0 & 0 \\ 1 & 0 \end{pmatrix}\begin{pmatrix} 1 \\ 0 \end{pmatrix} = \begin{pmatrix} 0 \\ 1 \end{pmatrix} \qquad \text{– decomposition equation}$$

p:

$$(0,1)\begin{pmatrix} 0 & 0 \\ 1 & 0 \end{pmatrix} = (1,0) \qquad \text{– composition equation}$$

$$\begin{pmatrix} 0 & 1 \\ 0 & 0 \end{pmatrix}\begin{pmatrix} 0 \\ 1 \end{pmatrix} = \begin{pmatrix} 1 \\ 0 \end{pmatrix} \qquad \text{– decomposition equation}$$

Absorption or emission of neutral π^0 does not modify the charge state of a nucleon, and can therefore be conveniently described as the identity operation, both for the composition and decomposition interactions.

Logic charges

The nucleon as a particle with two states, neutral and charged, can be mapped onto the basic binary alphabet of truth values

$$E_2 = \{ 0, 1 \} .$$

The fundamental alphabet of truth values

$$E_4 = \{ -1, 0, 1, 2 \}$$

requires a multiplet of higher symmetry, which includes the quadruplet of charged states with four different charges. The charge operator Q then takes the form:

$$Q = \begin{pmatrix} 1 & 0 & 0 & 0 \\ 0 & -1 & 0 & 0 \\ 0 & 0 & 0 & 0 \\ 0 & 0 & 0 & 2 \end{pmatrix},$$

where the diagonal elements representing the charges are in fact the joint set of eigenvalues of negation and tautology. The operator Q can be partitioned into two empty operators $\hat{0}$, and the diagonalized operators of negation and tautology.

The quadruplet represents an abstract quasi-particle whose logic status can decay in accordance with the laws of conservation.

From the eigenvectors of negation and tautology

$$| q_1 > = \begin{pmatrix} \frac{1}{2} \\ \frac{1}{2} \end{pmatrix} \qquad | q_2 > = \begin{pmatrix} \frac{1}{2} \\ -\frac{1}{2} \end{pmatrix}$$

we can construct 4 logic bispinors:

$$|A> = \begin{pmatrix} |q_1> \\ |q_1> \end{pmatrix} = \begin{pmatrix} \frac{1}{2} \\ \frac{1}{2} \\ \frac{1}{2} \\ \frac{1}{2} \end{pmatrix} = \frac{1}{2}\begin{pmatrix} 1 \\ 1 \\ 1 \\ 1 \end{pmatrix}$$

$$|B> = \begin{pmatrix} |q_2> \\ |q_2> \end{pmatrix} = \begin{pmatrix} \frac{1}{2} \\ -\frac{1}{2} \\ \frac{1}{2} \\ -\frac{1}{2} \end{pmatrix} = \frac{1}{2}\begin{pmatrix} 1 \\ -1 \\ 1 \\ -1 \end{pmatrix}$$

$$|C> = \begin{pmatrix} |q_1> \\ |q_2> \end{pmatrix} = \begin{pmatrix} \frac{1}{2} \\ \frac{1}{2} \\ \frac{1}{2} \\ -\frac{1}{2} \end{pmatrix} = \frac{1}{2}\begin{pmatrix} 1 \\ 1 \\ 1 \\ -1 \end{pmatrix}$$

$$|D> = \begin{pmatrix} |q_2> \\ |q_1> \end{pmatrix} = \begin{pmatrix} \frac{1}{2} \\ -\frac{1}{2} \\ \frac{1}{2} \\ \frac{1}{2} \end{pmatrix} = \frac{1}{2}\begin{pmatrix} 1 \\ -1 \\ 1 \\ 1 \end{pmatrix}$$

The bispinors form two orthogonal pairs

$$<A|B> = 0$$

$$<C|D> = 0 .$$

The charge operator Q operating on bispinors delivers 4 products:

$$Q|A> = \frac{1}{2}\begin{pmatrix} 1 \\ -1 \\ 0 \\ 2 \end{pmatrix}, \qquad \|Q|A>\| = 1$$

$$Q|B> = \frac{1}{2}\begin{pmatrix} 1 \\ 1 \\ 0 \\ -2 \end{pmatrix}, \qquad \|Q|B>\| = 0$$

$$Q\,|\,C> = \frac{1}{2}\begin{pmatrix} 1 \\ -1 \\ 0 \\ -2 \end{pmatrix}, \qquad \|Q\,|\,C>\| = -1$$

$$Q\,|\,D> = \frac{1}{2}\begin{pmatrix} 1 \\ 1 \\ 0 \\ 2 \end{pmatrix}, \qquad \|Q\,|\,D>\| = 2$$

The value of the charge is found as the logic norm of the product vectors obtained, that is, as the sums of their components that yield 4 different logic values

$$\{ -1, 0, 1, 2 \}$$

This is precisely the set of logic charges in the operator Q.

The fundamental alphabet of truth-values comprises two pairs with complementary truth status:

$$\overline{1} = 0$$

$$\overline{-1} = 2$$

This permits us to relate the purely logical concept of truth-value to the purely physical concept of bispinor. When the logic norms of the products of the charge operator Q with bispinors are complementary

$$\overline{\|Q|A>\|} = \|Q|B>\|$$

$$\overline{\|Q|C>\|} = \|Q|D>\|,$$

then the corresponding bispinors are orthogonal:

$$<A\,|\,B> = 0$$

$$<C\,|\,D> = 0$$

and vice versa.

Because a transition to the orthogonal bispinor modifies the logic charge of the product to a complementary charge, we are able to change, by purely physical transformation, the truth status of the system in question. This is yet another compelling reason for assuming that the laws of logic and consequently the fundamental laws of the mind have their roots in the fundamental domain. A synthesis of physical and logic notions implicit in fundamental physics becomes explicit in the matrix formulation of logic. Tout est logique!

3.5 Tensor logic

The computational reform accomplished in the matrix formulation of logic paves the way for the development of tensor logic, which is of great theoretical and applied significance. The problems of tensor logic are closely related to the fundamental problems of high-level intelligence and to the acute problems of parallel computing. An important property of a tensor is that, similar to a vector, it represents a covariant quantity. As the components of a tensor vary in different reference systems, with the passage from one system to another, they are transformed according to certain rules, so that the tensor contains the same information and represents the same quantity.

3.5.1 Hyperconnectives

Tensor logic is the n-dimensional generalization of standard 2-dimensional matrix logic. Whereas in standard logic the connective operators are generated as outer products of logic vectors,

$$|x\rangle\langle y| = L ,$$

in tensor logic it is the connective operators themselves which generate the connective operators of a higher order, the hyperspace operators H:

$$L_1 \otimes L_2 = H .$$

Tensor logic accommodates logic operations of a higher degree of abstraction and complexity and is ideally suited to account in a consistent and natural way for the parallelism inherent in intelligent operations.

Intelligent evaluations involve a great many variables which change and interact in a very complex manner. A key to deciphering this complexity lies in the massively parallel architecture of the human mind. In parallel logic a group of logic operators is handled simultaneously with definite rules relating the operators involved, very much in the same manner as the operations of the brain arise not only from massively parallel computations, but computations which are also bonded among themselves in an organized and congruent fashion. Tensor logic provides an excellent tool with which to study the parallel logic of the mind in terms of rigorous and clearly defined mathematical structures.

$\otimes$	I	$\urcorner$	$\wedge$	$\vee$	$\rightarrow$	$\leftarrow$	$\downarrow$	$\uparrow$	$\backslash$	$/$	$\perp$	$\top$	$\dashv$	$\vdash$	$\mathbf{1}$	$\mathbf{0}$
I	$\begin{pmatrix}\mathrm{I}&0\\0&\mathrm{I}\end{pmatrix}$	$\begin{pmatrix}\urcorner&0\\0&\urcorner\end{pmatrix}$	$\begin{pmatrix}\wedge&0\\0&\wedge\end{pmatrix}$	$\begin{pmatrix}\vee&0\\0&\vee\end{pmatrix}$	$\begin{pmatrix}\rightarrow&0\\0&\rightarrow\end{pmatrix}$	$\begin{pmatrix}\leftarrow&0\\0&\leftarrow\end{pmatrix}$	$\begin{pmatrix}\downarrow&0\\0&\downarrow\end{pmatrix}$	$\begin{pmatrix}\uparrow&0\\0&\uparrow\end{pmatrix}$	$\begin{pmatrix}\backslash&0\\0&\backslash\end{pmatrix}$	$\begin{pmatrix}/&0\\0&/\end{pmatrix}$	$\begin{pmatrix}\perp&0\\0&\perp\end{pmatrix}$	$\begin{pmatrix}\top&0\\0&\top\end{pmatrix}$	$\begin{pmatrix}\dashv&0\\0&\dashv\end{pmatrix}$	$\begin{pmatrix}\vdash&0\\0&\vdash\end{pmatrix}$	$\begin{pmatrix}1&0\\0&1\end{pmatrix}$	$(\mathbf{0})$
$\urcorner$	$\begin{pmatrix}0&\mathrm{I}\\\mathrm{I}&0\end{pmatrix}$	$\begin{pmatrix}0&\urcorner\\\urcorner&0\end{pmatrix}$	$\begin{pmatrix}0&\wedge\\\wedge&0\end{pmatrix}$	$\begin{pmatrix}0&\vee\\\vee&0\end{pmatrix}$	$\begin{pmatrix}0&\rightarrow\\\rightarrow&0\end{pmatrix}$	$\begin{pmatrix}0&\leftarrow\\\leftarrow&0\end{pmatrix}$	$\begin{pmatrix}0&\downarrow\\\downarrow&0\end{pmatrix}$	$\begin{pmatrix}0&\uparrow\\\uparrow&0\end{pmatrix}$	$\begin{pmatrix}0&\backslash\\\backslash&0\end{pmatrix}$	$\begin{pmatrix}0&/\\/&0\end{pmatrix}$	$\begin{pmatrix}0&\perp\\\perp&0\end{pmatrix}$	$\begin{pmatrix}0&\top\\\top&0\end{pmatrix}$	$\begin{pmatrix}0&\dashv\\\dashv&0\end{pmatrix}$	$\begin{pmatrix}0&\vdash\\\vdash&0\end{pmatrix}$	$\begin{pmatrix}0&1\\1&0\end{pmatrix}$	$(\mathbf{0})$
$\wedge$	$\begin{pmatrix}0&0\\0&\mathrm{I}\end{pmatrix}$	$\begin{pmatrix}0&0\\0&\urcorner\end{pmatrix}$	$\begin{pmatrix}0&0\\0&\wedge\end{pmatrix}$	$\begin{pmatrix}0&0\\0&\vee\end{pmatrix}$	$\begin{pmatrix}0&0\\0&\rightarrow\end{pmatrix}$	$\begin{pmatrix}0&0\\0&\leftarrow\end{pmatrix}$	$\begin{pmatrix}0&0\\0&\downarrow\end{pmatrix}$	$\begin{pmatrix}0&0\\0&\uparrow\end{pmatrix}$	$\begin{pmatrix}0&0\\0&\backslash\end{pmatrix}$	$\begin{pmatrix}0&0\\0&/\end{pmatrix}$	$\begin{pmatrix}0&0\\0&\perp\end{pmatrix}$	$\begin{pmatrix}0&0\\0&\top\end{pmatrix}$	$\begin{pmatrix}0&0\\0&\dashv\end{pmatrix}$	$\begin{pmatrix}0&0\\0&\vdash\end{pmatrix}$	$\begin{pmatrix}0&0\\0&1\end{pmatrix}$	$(\mathbf{0})$
$\vee$	$\begin{pmatrix}0&\mathrm{I}\\\mathrm{I}&\mathrm{I}\end{pmatrix}$	$\begin{pmatrix}0&\urcorner\\\urcorner&\urcorner\end{pmatrix}$	$\begin{pmatrix}0&\wedge\\\wedge&\wedge\end{pmatrix}$	$\begin{pmatrix}0&\vee\\\vee&\vee\end{pmatrix}$	$\begin{pmatrix}0&\rightarrow\\\rightarrow&\rightarrow\end{pmatrix}$	$\begin{pmatrix}0&\leftarrow\\\leftarrow&\leftarrow\end{pmatrix}$	$\begin{pmatrix}0&\downarrow\\\downarrow&\downarrow\end{pmatrix}$	$\begin{pmatrix}0&\uparrow\\\uparrow&\uparrow\end{pmatrix}$	$\begin{pmatrix}0&\backslash\\\backslash&\backslash\end{pmatrix}$	$\begin{pmatrix}0&/\\/&/\end{pmatrix}$	$\begin{pmatrix}0&\perp\\\perp&\perp\end{pmatrix}$	$\begin{pmatrix}0&\top\\\top&\top\end{pmatrix}$	$\begin{pmatrix}0&\dashv\\\dashv&\dashv\end{pmatrix}$	$\begin{pmatrix}0&\vdash\\\vdash&\vdash\end{pmatrix}$	$\begin{pmatrix}0&1\\1&1\end{pmatrix}$	$(\mathbf{0})$
$\rightarrow$	$\begin{pmatrix}\mathrm{I}&\mathrm{I}\\0&\mathrm{I}\end{pmatrix}$	$\begin{pmatrix}\urcorner&\urcorner\\0&\urcorner\end{pmatrix}$	$\begin{pmatrix}\wedge&\wedge\\0&\wedge\end{pmatrix}$	$\begin{pmatrix}\vee&\vee\\0&\vee\end{pmatrix}$	$\begin{pmatrix}\rightarrow&\rightarrow\\0&\rightarrow\end{pmatrix}$	$\begin{pmatrix}\leftarrow&\leftarrow\\0&\leftarrow\end{pmatrix}$	$\begin{pmatrix}\downarrow&\downarrow\\0&\downarrow\end{pmatrix}$	$\begin{pmatrix}\uparrow&\uparrow\\0&\uparrow\end{pmatrix}$	$\begin{pmatrix}\backslash&\backslash\\0&\backslash\end{pmatrix}$	$\begin{pmatrix}/&/\\0&/\end{pmatrix}$	$\begin{pmatrix}\perp&\perp\\0&\perp\end{pmatrix}$	$\begin{pmatrix}\top&\top\\0&\top\end{pmatrix}$	$\begin{pmatrix}\dashv&\dashv\\0&\dashv\end{pmatrix}$	$\begin{pmatrix}\vdash&\vdash\\0&\vdash\end{pmatrix}$	$\begin{pmatrix}1&1\\0&1\end{pmatrix}$	$(\mathbf{0})$
$\leftarrow$	$\begin{pmatrix}\mathrm{I}&0\\\mathrm{I}&\mathrm{I}\end{pmatrix}$	$\begin{pmatrix}\urcorner&0\\\urcorner&\urcorner\end{pmatrix}$	$\begin{pmatrix}\wedge&0\\\wedge&\wedge\end{pmatrix}$	$\begin{pmatrix}\vee&0\\\vee&\vee\end{pmatrix}$	$\begin{pmatrix}\rightarrow&0\\\rightarrow&\rightarrow\end{pmatrix}$	$\begin{pmatrix}\leftarrow&0\\\leftarrow&\leftarrow\end{pmatrix}$	$\begin{pmatrix}\downarrow&0\\\downarrow&\downarrow\end{pmatrix}$	$\begin{pmatrix}\uparrow&0\\\uparrow&\uparrow\end{pmatrix}$	$\begin{pmatrix}\backslash&0\\\backslash&\backslash\end{pmatrix}$	$\begin{pmatrix}/&0\\/&/\end{pmatrix}$	$\begin{pmatrix}\perp&0\\\perp&\perp\end{pmatrix}$	$\begin{pmatrix}\top&0\\\top&\top\end{pmatrix}$	$\begin{pmatrix}\dashv&0\\\dashv&\dashv\end{pmatrix}$	$\begin{pmatrix}\vdash&0\\\vdash&\vdash\end{pmatrix}$	$\begin{pmatrix}1&0\\1&1\end{pmatrix}$	$(\mathbf{0})$
$\downarrow$	$\begin{pmatrix}\mathrm{I}&\mathrm{I}\\\mathrm{I}&0\end{pmatrix}$	$\begin{pmatrix}\urcorner&\urcorner\\\urcorner&0\end{pmatrix}$	$\begin{pmatrix}\wedge&\wedge\\\wedge&0\end{pmatrix}$	$\begin{pmatrix}\vee&\vee\\\vee&0\end{pmatrix}$	$\begin{pmatrix}\rightarrow&\rightarrow\\\rightarrow&0\end{pmatrix}$	$\begin{pmatrix}\leftarrow&\leftarrow\\\leftarrow&0\end{pmatrix}$	$\begin{pmatrix}\downarrow&\downarrow\\\downarrow&0\end{pmatrix}$	$\begin{pmatrix}\uparrow&\uparrow\\\uparrow&0\end{pmatrix}$	$\begin{pmatrix}\backslash&\backslash\\\backslash&0\end{pmatrix}$	$\begin{pmatrix}/&/\\/&0\end{pmatrix}$	$\begin{pmatrix}\perp&\perp\\\perp&0\end{pmatrix}$	$\begin{pmatrix}\top&\top\\\top&0\end{pmatrix}$	$\begin{pmatrix}\dashv&\dashv\\\dashv&0\end{pmatrix}$	$\begin{pmatrix}\vdash&\vdash\\\vdash&0\end{pmatrix}$	$\begin{pmatrix}1&1\\1&0\end{pmatrix}$	$(\mathbf{0})$
$\uparrow$	$\begin{pmatrix}\mathrm{I}&0\\0&0\end{pmatrix}$	$\begin{pmatrix}\urcorner&0\\0&0\end{pmatrix}$	$\begin{pmatrix}\wedge&0\\0&0\end{pmatrix}$	$\begin{pmatrix}\vee&0\\0&0\end{pmatrix}$	$\begin{pmatrix}\rightarrow&0\\0&0\end{pmatrix}$	$\begin{pmatrix}\leftarrow&0\\0&0\end{pmatrix}$	$\begin{pmatrix}\downarrow&0\\0&0\end{pmatrix}$	$\begin{pmatrix}\uparrow&0\\0&0\end{pmatrix}$	$\begin{pmatrix}\backslash&0\\0&0\end{pmatrix}$	$\begin{pmatrix}/&0\\0&0\end{pmatrix}$	$\begin{pmatrix}\perp&0\\0&0\end{pmatrix}$	$\begin{pmatrix}\top&0\\0&0\end{pmatrix}$	$\begin{pmatrix}\dashv&0\\0&0\end{pmatrix}$	$\begin{pmatrix}\vdash&0\\0&0\end{pmatrix}$	$\begin{pmatrix}1&0\\0&0\end{pmatrix}$	$(\mathbf{0})$
$\backslash$	$\begin{pmatrix}0&0\\\mathrm{I}&0\end{pmatrix}$	$\begin{pmatrix}0&0\\\urcorner&0\end{pmatrix}$	$\begin{pmatrix}0&0\\\wedge&0\end{pmatrix}$	$\begin{pmatrix}0&0\\\vee&0\end{pmatrix}$	$\begin{pmatrix}0&0\\\rightarrow&0\end{pmatrix}$	$\begin{pmatrix}0&0\\\leftarrow&0\end{pmatrix}$	$\begin{pmatrix}0&0\\\downarrow&0\end{pmatrix}$	$\begin{pmatrix}0&0\\\uparrow&0\end{pmatrix}$	$\begin{pmatrix}0&0\\\backslash&0\end{pmatrix}$	$\begin{pmatrix}0&0\\/&0\end{pmatrix}$	$\begin{pmatrix}0&0\\\perp&0\end{pmatrix}$	$\begin{pmatrix}0&0\\\top&0\end{pmatrix}$	$\begin{pmatrix}0&0\\\dashv&0\end{pmatrix}$	$\begin{pmatrix}0&0\\\vdash&0\end{pmatrix}$	$\begin{pmatrix}0&0\\1&0\end{pmatrix}$	$(\mathbf{0})$
$/$	$\begin{pmatrix}0&\mathrm{I}\\0&0\end{pmatrix}$	$\begin{pmatrix}0&\urcorner\\0&0\end{pmatrix}$	$\begin{pmatrix}0&\wedge\\0&0\end{pmatrix}$	$\begin{pmatrix}0&\vee\\0&0\end{pmatrix}$	$\begin{pmatrix}0&\rightarrow\\0&0\end{pmatrix}$	$\begin{pmatrix}0&\leftarrow\\0&0\end{pmatrix}$	$\begin{pmatrix}0&\downarrow\\0&0\end{pmatrix}$	$\begin{pmatrix}0&\uparrow\\0&0\end{pmatrix}$	$\begin{pmatrix}0&\backslash\\0&0\end{pmatrix}$	$\begin{pmatrix}0&/\\0&0\end{pmatrix}$	$\begin{pmatrix}0&\perp\\0&0\end{pmatrix}$	$\begin{pmatrix}0&\top\\0&0\end{pmatrix}$	$\begin{pmatrix}0&\dashv\\0&0\end{pmatrix}$	$\begin{pmatrix}0&\vdash\\0&0\end{pmatrix}$	$\begin{pmatrix}0&1\\0&0\end{pmatrix}$	$(\mathbf{0})$
$\perp$	$\begin{pmatrix}0&0\\\mathrm{I}&\mathrm{I}\end{pmatrix}$	$\begin{pmatrix}0&0\\\urcorner&\urcorner\end{pmatrix}$	$\begin{pmatrix}0&0\\\wedge&\wedge\end{pmatrix}$	$\begin{pmatrix}0&0\\\vee&\vee\end{pmatrix}$	$\begin{pmatrix}0&0\\\rightarrow&\rightarrow\end{pmatrix}$	$\begin{pmatrix}0&0\\\leftarrow&\leftarrow\end{pmatrix}$	$\begin{pmatrix}0&0\\\downarrow&\downarrow\end{pmatrix}$	$\begin{pmatrix}0&0\\\uparrow&\uparrow\end{pmatrix}$	$\begin{pmatrix}0&0\\\backslash&\backslash\end{pmatrix}$	$\begin{pmatrix}0&0\\/&/\end{pmatrix}$	$\begin{pmatrix}0&0\\\perp&\perp\end{pmatrix}$	$\begin{pmatrix}0&0\\\top&\top\end{pmatrix}$	$\begin{pmatrix}0&0\\\dashv&\dashv\end{pmatrix}$	$\begin{pmatrix}0&0\\\vdash&\vdash\end{pmatrix}$	$\begin{pmatrix}0&0\\1&1\end{pmatrix}$	$(\mathbf{0})$
$\top$	$\begin{pmatrix}\mathrm{I}&\mathrm{I}\\0&0\end{pmatrix}$	$\begin{pmatrix}\urcorner&\urcorner\\0&0\end{pmatrix}$	$\begin{pmatrix}\wedge&\wedge\\0&0\end{pmatrix}$	$\begin{pmatrix}\vee&\vee\\0&0\end{pmatrix}$	$\begin{pmatrix}\rightarrow&\rightarrow\\0&0\end{pmatrix}$	$\begin{pmatrix}\leftarrow&\leftarrow\\0&0\end{pmatrix}$	$\begin{pmatrix}\downarrow&\downarrow\\0&0\end{pmatrix}$	$\begin{pmatrix}\uparrow&\uparrow\\0&0\end{pmatrix}$	$\begin{pmatrix}\backslash&\backslash\\0&0\end{pmatrix}$	$\begin{pmatrix}/&/\\0&0\end{pmatrix}$	$\begin{pmatrix}\perp&\perp\\0&0\end{pmatrix}$	$\begin{pmatrix}\top&\top\\0&0\end{pmatrix}$	$\begin{pmatrix}\dashv&\dashv\\0&0\end{pmatrix}$	$\begin{pmatrix}\vdash&\vdash\\0&0\end{pmatrix}$	$\begin{pmatrix}1&1\\0&0\end{pmatrix}$	$(\mathbf{0})$
$\dashv$	$\begin{pmatrix}0&\mathrm{I}\\0&\mathrm{I}\end{pmatrix}$	$\begin{pmatrix}0&\urcorner\\0&\urcorner\end{pmatrix}$	$\begin{pmatrix}0&\wedge\\0&\wedge\end{pmatrix}$	$\begin{pmatrix}0&\vee\\0&\vee\end{pmatrix}$	$\begin{pmatrix}0&\rightarrow\\0&\rightarrow\end{pmatrix}$	$\begin{pmatrix}0&\leftarrow\\0&\leftarrow\end{pmatrix}$	$\begin{pmatrix}0&\downarrow\\0&\downarrow\end{pmatrix}$	$\begin{pmatrix}0&\uparrow\\0&\uparrow\end{pmatrix}$	$\begin{pmatrix}0&\backslash\\0&\backslash\end{pmatrix}$	$\begin{pmatrix}0&/\\0&/\end{pmatrix}$	$\begin{pmatrix}0&\perp\\0&\perp\end{pmatrix}$	$\begin{pmatrix}0&\top\\0&\top\end{pmatrix}$	$\begin{pmatrix}0&\dashv\\0&\dashv\end{pmatrix}$	$\begin{pmatrix}0&\vdash\\0&\vdash\end{pmatrix}$	$\begin{pmatrix}0&1\\0&1\end{pmatrix}$	$(\mathbf{0})$
$\vdash$	$\begin{pmatrix}\mathrm{I}&0\\\mathrm{I}&0\end{pmatrix}$	$\begin{pmatrix}\urcorner&0\\\urcorner&0\end{pmatrix}$	$\begin{pmatrix}\wedge&0\\\wedge&0\end{pmatrix}$	$\begin{pmatrix}\vee&0\\\vee&0\end{pmatrix}$	$\begin{pmatrix}\rightarrow&0\\\rightarrow&0\end{pmatrix}$	$\begin{pmatrix}\leftarrow&0\\\leftarrow&0\end{pmatrix}$	$\begin{pmatrix}\downarrow&0\\\downarrow&0\end{pmatrix}$	$\begin{pmatrix}\uparrow&0\\\uparrow&0\end{pmatrix}$	$\begin{pmatrix}\backslash&0\\\backslash&0\end{pmatrix}$	$\begin{pmatrix}/&0\\/&0\end{pmatrix}$	$\begin{pmatrix}\perp&0\\\perp&0\end{pmatrix}$	$\begin{pmatrix}\top&0\\\top&0\end{pmatrix}$	$\begin{pmatrix}\dashv&0\\\dashv&0\end{pmatrix}$	$\begin{pmatrix}\vdash&0\\\vdash&0\end{pmatrix}$	$\begin{pmatrix}1&0\\1&0\end{pmatrix}$	$(\mathbf{0})$
$\mathbf{1}$	$\begin{pmatrix}\mathrm{I}&\mathrm{I}\\\mathrm{I}&\mathrm{I}\end{pmatrix}$	$\begin{pmatrix}\urcorner&\urcorner\\\urcorner&\urcorner\end{pmatrix}$	$\begin{pmatrix}\wedge&\wedge\\\wedge&\wedge\end{pmatrix}$	$\begin{pmatrix}\vee&\vee\\\vee&\vee\end{pmatrix}$	$\begin{pmatrix}\rightarrow&\rightarrow\\\rightarrow&\rightarrow\end{pmatrix}$	$\begin{pmatrix}\leftarrow&\leftarrow\\\leftarrow&\leftarrow\end{pmatrix}$	$\begin{pmatrix}\downarrow&\downarrow\\\downarrow&\downarrow\end{pmatrix}$	$\begin{pmatrix}\uparrow&\uparrow\\\uparrow&\uparrow\end{pmatrix}$	$\begin{pmatrix}\backslash&\backslash\\\backslash&\backslash\end{pmatrix}$	$\begin{pmatrix}/&/\\/&/\end{pmatrix}$	$\begin{pmatrix}\perp&\perp\\\perp&\perp\end{pmatrix}$	$\begin{pmatrix}\top&\top\\\top&\top\end{pmatrix}$	$\begin{pmatrix}\dashv&\dashv\\\dashv&\dashv\end{pmatrix}$	$\begin{pmatrix}\vdash&\vdash\\\vdash&\vdash\end{pmatrix}$	$\begin{pmatrix}1&1\\1&1\end{pmatrix}$	$(\mathbf{0})$
$\mathbf{0}$	$(\mathbf{0})$	$(\mathbf{0})$	$(\mathbf{0})$	$(\mathbf{0})$	$(\mathbf{0})$	$(\mathbf{0})$	$(\mathbf{0})$	$(\mathbf{0})$	$(\mathbf{0})$	$(\mathbf{0})$	$(\mathbf{0})$	$(\mathbf{0})$	$(\mathbf{0})$	$(\mathbf{0})$	$(\mathbf{0})$	$(\mathbf{0})$

Table 3.7
Tensor logic multiplication table

Operators

In the hyperspace extension of logic, constructs appear that are more familiar to physicists or mathematicians than to traditional logicians. Tensor logic is defined in n-dimensional logic spaces of the bra and ket logic vectors $\langle h|$ and $|h\rangle$, which are extended in the orthonormal basis:

$$\begin{matrix} (1 & 0 & 0 & \dots & 0) \\ (0 & 1 & 0 & \dots & 0) \\ (0 & 0 & 1 & \dots & 0) \\ . & . & . & \dots & . \\ (0 & 0 & 0 & \dots & 1) \end{matrix}$$

The 2-dimensional bra and ket logic vectors which we have considered so far thus represent the simplest possible set of logic vectors. The hyperspace vectors satisfy the condition of logic normalization for the vector components,

$$\sum_{i}^{n} h_i = 1 .$$

Consequently, the logic norm of the outer products of the bra and ket vectors is a tautology:

$$\left\| |h_i\rangle\langle h_j| \right\| = \|\Omega\| \equiv 1 ,$$

and

$$0 \leq \langle h_i | H | h_j \rangle \leq 1.$$

The hyperspace logic operators are hyperconnectives, which can be viewed as the matrix juxtaposition of the basic logic connectives. Just as hyperspace logic vectors are an n-dimensional generalization of 2-dimensional ones, the hyperspace logic operators are an $(n \times n)$ generalization of (2×2) matrix logic operators. The formation of a hyperconnective signifies the formation of ordered links between constituent operators, which gives each of them a specific place in the hyperspace operator.

In conventional terms a hyperconnective can be interpreted simply as a specially ordered truth-table for a compound logic function. In contrast, in matrix logic terms it represents a mathematical object to which we can apply various matrix and tensor operations, extracting useful data and dealing with different outcomes of logic functions simultaneously, in parallel fashion.

Consider an (8 × 8) hyperspace logic operator comprising 16 standard logic operators L_{ij}:

$$H = \begin{pmatrix} \begin{pmatrix} (L_{11}) & (L_{12}) \\ (L_{21}) & (L_{22}) \end{pmatrix} & \begin{pmatrix} (L_{13}) & (L_{14}) \\ (L_{23}) & (L_{24}) \end{pmatrix} \\ \begin{pmatrix} (L_{31}) & (L_{32}) \\ (L_{41}) & (L_{42}) \end{pmatrix} & \begin{pmatrix} (L_{33}) & (L_{34}) \\ (L_{43}) & (L_{44}) \end{pmatrix} \end{pmatrix}$$

The H-operator is a square matrix which can be partitioned into a set of (n/2 × n/2) suboperators, which could in their own turn be partitioned into a set of (n/4 × n/4) suboperators, and so on until the H-operator is partitioned into a set of basic (2 × 2) logic operators. If n runs to infinity, the above reduction takes the form of an infinite regress.
The complex structure of logic connectives, already demonstrated for standard matrix logic, is further elaborated in tensor logic. The range of complexification of logic operations made possible by tensor logic is practically limitless. The extremely difficult problems of parallelism are placed on a solid mathematical foundation.

Since the results presented here are applicable to a logic of any finite order, it will be sufficient for our purposes if we consider (4 × 4) hyperspace operators

$$H = \begin{pmatrix} L_{11} & L_{12} \\ L_{21} & L_{22} \end{pmatrix}, \quad L_{ij} \in \Omega^+$$

Even a slight jump from 2-D logic space to a 4-D space is accompanied by a huge increase in the possible number of logic connective operators. There are $2^{16} = 65536$ different (4 × 4) hyperspace logic operators, enough to accommodate the most elaborate intelligence. With further increases in logic dimensions the number of operators explodes exponentially, reaching astronomical magnitudes.

Matrix product

For the matrix product of hyperspace operators, H_1H_2,

$$\begin{aligned} H_1 &= L_1 \otimes L_2 \\ H_2 &= L_3 \otimes L_4 \end{aligned}$$

the following fundamental equality holds:

$$H_1H_2 = (L_1 \otimes L_2)(L_3 \otimes L_4) = L_1L_3 \otimes L_2L_4 .$$

This is established as follows:

$$\begin{aligned} &< xx' \mid (L_1 \otimes L_2)(L_3 \otimes L_4) \mid yy'> = \\ &= \sum_{zz'} < xx' \mid L_1 \otimes L_2 \mid zz'> < zz' \mid L_3 \otimes L_4 \mid yy'> = \\ &= \sum_{z} < x \mid L_1 \mid z > < z \mid L_3 \mid y > \cdot \sum_{z'} < x' \mid L_2 \mid z'> < z' \mid L_4 \mid y'> = \\ &= < x \mid L_1L_3 \mid y > < x' \mid L_2L_4 \mid y'> = < xx' \mid L_1L_3 \otimes L_2L_4 \mid yy'> . \end{aligned}$$

In the general case we have:

$$\begin{aligned} &(L_1 \otimes L_2 \otimes \ldots \otimes L_n)(L_{n+1} \otimes L_{n+2} \otimes \ldots \otimes L_m) = \\ &= L_1L_{n+1} \otimes L_2L_{n+2} \otimes \ldots \otimes L_nL_m . \end{aligned}$$

Given a matrix logic string in the form of a fundamental product, we can contract the latter into the corresponding tensor logic product:

$$<x \mid L_1 \mid y> <x' \mid L_2 \mid y'> = <xx' \mid L_1 \otimes L_2 \mid yy'> = <xx' \mid H \mid yy'> ,$$

where L_1, L_2 are (2×2) logic operators and H is a (4×4) logic operator. By induction we establish the general rule for the contraction of a fundamental product of any complexity:

$$\begin{aligned} &<x \mid L_1 \mid y> <x' \mid L_2 \mid y'> \ldots <x'' \mid L_m \mid y''> = \\ &= <xx' \ldots x'' \mid L_1 \otimes L_2 \otimes \ldots \otimes L_m \mid yy' \ldots y''> = \\ &= <h_i \mid H \mid h_j> , \end{aligned}$$

where $L_1, L_2, \ldots, L_n$ are 2-dimensional logic operators, $H = L_1 \otimes L_2 \otimes \ldots \otimes L_m$ is a 2^m-dimensional hyperspace operator, and $<h_i \mid = <xx' \ldots x'' \mid$ and $\mid h_j> = \mid yy' \ldots y''>$ are hyperspace logic vectors.

A fundamental product comprising m atomic products defines a 2^m-dimensional hyperspace. Conversely, the dimension of a hyperspace defines the number of atomic products in the corresponding fundamental product.

Tensor logic, by virtue of the organization of the hyperspace operators, is characterized by parallel architecture. A fundamental product comprising a sequence of bracket products is contracted into a single hyperspace bracket product. Quite naturally, when a logic problem has both sequential and parallel solutions, it is desirable that we should be able to pass from the sequential code to the parallel code and vice versa. This establishes two interrelated problems in hyperspace logic: the direct problem of the contraction of a fundamental product,

$$\langle x \mid L_1 \mid y\rangle \langle x' \mid L_2 \mid y'\rangle \ldots \;\Rightarrow\; \langle xx'\ldots \mid H \mid yy'\ldots\rangle ,$$

and the inverse problem of the decomposition of a hyperspace bracket product into the corresponding fundamental product:

$$\langle xx'\ldots \mid H \mid yy'\ldots\rangle \;\Rightarrow\; \langle x \mid L_1 \mid y\rangle \langle x' \mid L_2 \mid y'\rangle \ldots$$

When a fundamental product is the algebraic sum of products of bracket products, a passage to hyperspace logic may very well produce an algebraic sum of hyperspace bracket products of different dimensions.

It is important to emphasize that while the direct problem clearly always has a solution, for the inverse problem it may occur that the hyperspace logic operator cannot be presented as a tensor product of standard logic operators. The fact that hyperspace logic is not automatically broken down into a standard matrix string signifies a special role for hyperspace logic operations.

3.5.2 Fractal logic operators

An important subset of hyperspace logic operators is represented by the fractal operators F, obtained as Kronecker powers of logic operators:

$$F = L^{[2]} = L \otimes L .$$

The fractal operator is thus an exact replica or scaled model of itself. In the tensor logic multiplication table fractal operators reside on the diagonal. From the fundamental equality,

$$(L \otimes L)\,(L \otimes L) = LL \otimes LL ,$$

we deduce that for fractals the Kronecker power and matrix power commute:

$$(L^{[2]})^2 = (L^2)^{[2]}$$

In addition to fractal operators we define an important associated concept of antifractals:

$$L \otimes \bar{L},$$

that is, a tensor product of a logic operator and its complement.

The antifractals $L \otimes \bar{L}$ are characteristic of the hyperspace contraction of transmultiplication tautologies and contradictions. By transmultiplication fusion we obtain a contradiction:

$$\langle x | L | y \rangle \langle x | \bar{L} | y \rangle = \langle x | L \ast \bar{L} | y \rangle = \langle x | \hat{0} | y \rangle .$$

The corresponding tensor contradiction is an antifractal product:

$$\langle x | L | y \rangle \langle x | \bar{L} | y \rangle = \langle xx | L \otimes \bar{L} | yy \rangle .$$

Thus in tensor logic the transmultiplication tautologies and contradictions can be recognized by the antifractal structure of the corresponding hyperspace operators:

$$\langle h_i | L \otimes \bar{L} | h_j \rangle \quad \text{– contradiction}$$

$$\overline{\langle h_i | L \otimes \bar{L} | h_j \rangle} \quad \text{– tautology} .$$

The tensors of tautology and contradiction are invariant in all reference systems, that is, valid under any interpretation. A change of interpretation is viewed as a rotation of logic space, leading to a transition from one basis to another.

Absorption

If the absorption law holds for a fundamental product,

$$\langle x | L \ast L | y \rangle = \langle x | L | y \rangle ,$$

it also holds for a corresponding fractal. The fractal can then be convoluted to a space with fewer dimensions:

$$\langle xx | L \otimes L | yy \rangle = \langle x | L | y \rangle .$$

The idempotence of conventional logic finds in the fractal operations a metric analogue of a higher order.

Whereas there are no systematic links between ordinary and transmultiplication logic products, a tensor product and a transmultiplication product are closely related: the latter is contained in the former as a suboperator. In 4-D logic space the product $L_1 \ast L_2$ is obtained from $L_1 \otimes L_2$ by deletion of a grid of certain rows, 2 and 3, and certain columns, 2 and 3:

$$L_1 \otimes L_2 = \begin{pmatrix} \otimes & \times & \times & \otimes \\ \times & \times & \times & \times \\ \times & \times & \times & \times \\ \otimes & \times & \times & \otimes \end{pmatrix} \Rightarrow L_1 \ast L_2 = \begin{pmatrix} \otimes & \otimes \\ \otimes & \otimes \end{pmatrix}$$

The general selection rule is easily formulated by induction.

3.5.3 Logic subspaces

In tensor logic we construct the negation operator as the juxtaposition of the standard negation operators along the secondary diagonal. For 4-dimensional logic space we obtain

$$\neg_4 = \begin{pmatrix} \hat{0} & \neg_2 \\ \neg_2 & \hat{0} \end{pmatrix} = \begin{pmatrix} 0 & 0 & 0 & 1 \\ 0 & 0 & 1 & 0 \\ 0 & 1 & 0 & 0 \\ 1 & 0 & 0 & 0 \end{pmatrix}$$

$$\neg_4^2 = I_4$$

where I_4 is a 4-D identity and the subscripts 4 and 2 indicate the dimensionality of the logic space.

The orthogonal transform of a hyperspace operator H in a 4-dimensional logic space

$$\theta H = \theta (L_1 \otimes L_2)$$

is a tensor product of the orthogonal transforms in a 2-dimensional logic space:

$$\theta (L_1 \otimes L_2) = \theta L_1 \otimes \theta L_2$$

Explicitly:

$$\neg_4 (L_1 \otimes L_2) \neg_4 = (\neg_2 L_1 \neg_2) \otimes (\neg_2 L_2 \neg_2)$$

Thus, the orthogonal transform θ acts with respect to the tensor product as a regressing linear operator:

$$\theta\,(L_1 \otimes L_2) = \theta\, L_1 \otimes \theta\, L_2$$

In fact θ comprises two regressing linear operators:

$$\urcorner_4\,(L_1 \otimes L_2) = \urcorner_2\, L_1 \otimes \urcorner_2\, L_2$$

and

$$(L_1 \otimes L_2)\,\urcorner_4 = L_1\,\urcorner_2 \otimes L_2\,\urcorner_2\,.$$

It is important to emphasize that the dimensionality of the negation operator regresses from 4 to 2 as we pass from the left side of the equation to the right side. The possibility of expressing logic operations in n-D space via logic operations in (n–2)-D logic space is an important feature of tensor logic with clear practical applications. The transition from hyperspace to two subspaces of fewer dimensions, that is, of lesser complexity, significantly simplifies the solution of logic problems.

When the orthogonal transformation is applied to a hyperspace bracket product, just as in the standard logic space, we can either apply it to an operator or direct it towards the adjoint logic vectors:

$$\langle x_1\, x_2\, x_3\, x_4 \mid \theta\,(L_1 \otimes L_2) \mid y_1\, y_2\, y_3\, y_4\rangle =$$

$$= \langle x_1\, x_2\, x_3\, x_4 \mid \theta\, L_1 \otimes \theta\, L_2 \mid y_1\, y_2\, y_3\, y_4\rangle =$$

$$= \langle x_4\, x_3\, x_2\, x_1 \mid L_1 \otimes L_2 \mid y_4\, y_3\, y_2\, y_1\rangle\,.$$

The hyperspace negation operator performs a cyclic shift of the logic vector components. In this shift the lower pair of components is recombined with the upper pair, while components within each pair are recombined as well.

The action of the negation operator in hyperspace is of a more complex nature than in 2-D logic space. While in the standard logic space the negation operation is also identical to complementation,

$$\urcorner \mid p\rangle = \mid \bar{p}\rangle\,,$$

in tensor logic the product of negation and a vector is generally not equal to the complementation of that vector.

As with the orthogonal transformation, the matrix logic operations of transposition and contra-position are also linear operators in tensor logic:

$$\widetilde{(L_1 \otimes L_2)} = \tilde{L}_1 \otimes \tilde{L}_2$$

$$\overbracket{(L_1 \otimes L_2)} = \overbracket{L_1} \otimes \overbracket{L_2}$$

Compare this with the inversion occurring in standard logic:

$$\widetilde{L_1 L_2} = \tilde{L}_2\, \tilde{L}_1$$

$$\overbracket{L_1\ L_2} = \overbracket{L_2}\ \overbracket{L_1}$$

Unlike in ordinary matrix products, there is no reversal of the order in tensor logic.
From the general formula we derive particular rules

for contrasymmetric operators:

$$\theta\,(L_1 \otimes L_2) = \tilde{L}_1 \otimes \tilde{L}_2 = \widetilde{(L_1 \otimes L_2)}$$

for symmetric operators:

$$\theta\,(L_1 \otimes L_2) = \overbracket{L_1} \otimes \overbracket{L_2} = \overbracket{(L_1 \otimes L_2)}$$

for bi-asymmetric operators:

$$\theta\,(L_1 \otimes L_2) = \overline{L_1} \otimes \overline{L_2} \neq \overline{L_1 \otimes L_2}$$

for bisymmetric operators:

$$\theta\,(L_1 \otimes L_2) = L_1 \otimes L_2 .$$

For a tensor product the inversion, typical of matrix operations, is cancelled. This also occurs with the inverse of a hyperspace operator:

$$H^{-1} = (L_1 \otimes L_2)^{-1} = L_1^{-1} \otimes L_2^{-1} ,$$

whereas in 2–D logic

$$(L_1 L_2)^{-1} = L_2^{-1} L_1^{-1} .$$

Earlier we determined and examined in detail the non-commutativity of matrix logic operators. Tensor logic is also non-commuting, as in general the logic operators in a tensor product do not commute:

$$L_1 \otimes L_2 \neq L_2 \otimes L_1$$

However, with second and third column transposition and second and third row transposition, $L_2 \otimes L_1$ is changed into $L_1 \otimes L_2$. This result applies in general: the elements of $H_2 \otimes H_1$ are simply a rearrangement of those of $H_1 \otimes H_2$, and one product can be translated into the other by the appropriate row and column transpositions. Thus the limitations of non-commutativity encountered with ordinary matrix products of logic operators have, to a certain extent, been overcome.

From the general covariance rule

$$L_1 \otimes L_2 = \theta \overline{\tilde{L}_1} \otimes \theta \overline{\tilde{L}_2}$$

we derive the following sequence of results:

$$L_1 \otimes L_2 = \theta \left(\overline{\tilde{L}_1} \otimes \overline{\tilde{L}_2} \right)$$

$$\theta (L_1 \otimes L_2) = \overline{\tilde{L}_1} \otimes \overline{\tilde{L}_2}$$

$$\theta (L_1 \otimes L_2) = \overline{\tilde{L}_1 \otimes \tilde{L}_2}$$

$$\theta (\overline{L_1 \otimes L_2}) = \tilde{L}_1 \otimes \tilde{L}_2$$

$$\theta (\overline{L_1 \otimes L_2}) = \widetilde{L_1 \otimes L_2}$$

$$\theta (\widetilde{\overline{L_1 \otimes L_2}}) = L_1 \otimes L_2$$

Once again we have been able to express the operation of 4-D logic in terms of the standard operations of 2-D logic. This important principle of separation of a logic space into two interacting subspaces may hold the key to the fundamental division of the brain into left and right brains:

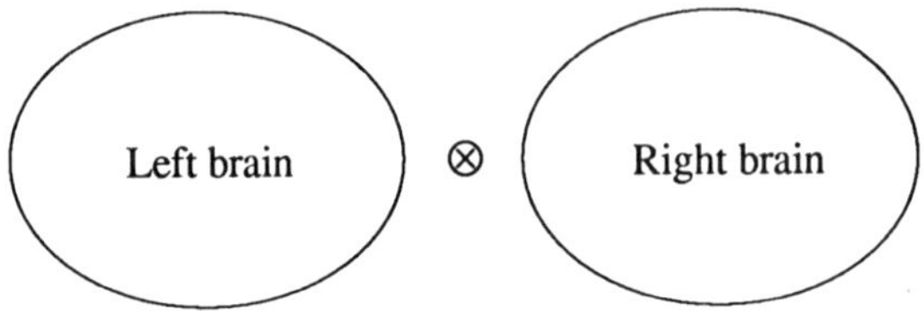

The global operator of the brain arises from the left and right brain tensor and transmultiplication interactions. Because a subspace of a logic space is a subset which itself is a logic space, each of the subbrains is fully capable of operating on its own. The opportunities for experimental logic are not as remote as one may think.

CHAPTER 4

PROBLEMS AND SOLUTIONS

Problem

Investigate the rotation of a logic space which translates a logic vector $| p>$ into the complementary vector $| \bar{p}>$. How are the complementary vectors located with respect to the undecided axis of the logic coordinate system?

Solution

The complementary logic vectors are located symmetrically with respect to the undecided axis:

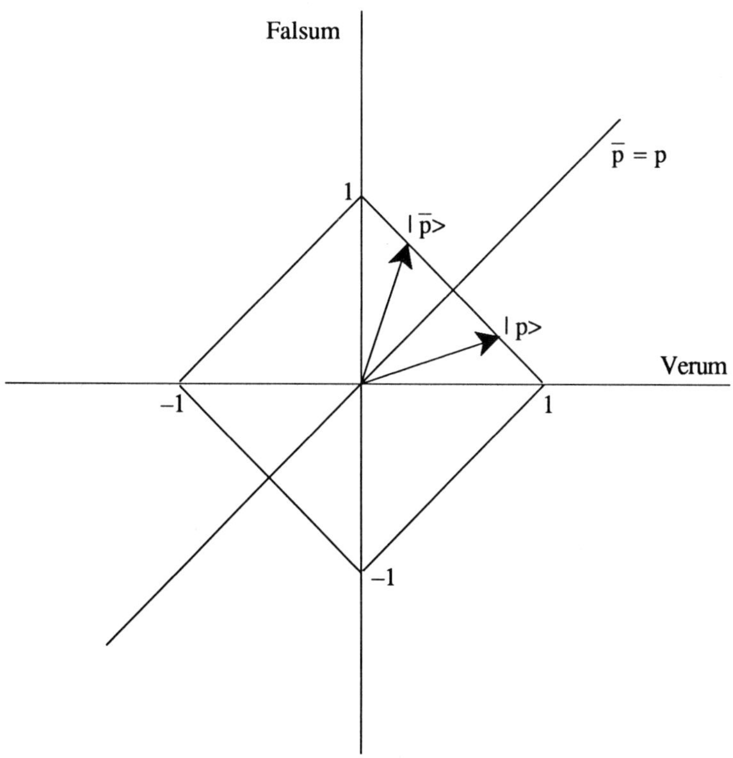

Problem

How many different metric intervals can be generated in logic space by the atomic products $<p\,|\,L\,|\,p>$?

Solution

Isotropic intervals:

$$<p\,|\,\blacktriangledown\,|\,p>,\quad <p\,|\,\blacktriangle\,|\,p>,\quad <p\,|\,\hat{0}\,|\,p>.$$

Euclidian intervals:

$$\pm(\bar{p}^2 + p^2),\quad \pm\bar{p}^2,\ \pm p^2.$$

Pseudo-Euclidian interval:

$$\pm(\bar{p}^2 - p^2).$$

Riemannian intervals:

$$\pm\bar{p}p,\ \pm 2\bar{p}p,\ \pm(p^2 + \bar{p}p),\ \pm(p^2 - \bar{p}p),\ \pm(\bar{p}^2 + \bar{p}p),\ \pm(\bar{p}^2 - \bar{p}p),$$
$$\pm(\bar{p}^2 + 2\bar{p}p),\ \pm(\bar{p}^2 - 2\bar{p}p),\ \pm(\bar{p}^2 + \bar{p}p + p^2),\ \pm(\bar{p}^2 - \bar{p}p - p^2),$$
$$\pm(\bar{p}^2 - \bar{p}p + p^2),\ \pm(\bar{p}^2 + \bar{p}p - p^2),\ \pm(2\bar{p}p + p^2),\ \pm(2\bar{p}p - p^2),$$
$$\pm(\bar{p}^2 - 2\bar{p}p - p^2),\ \pm(\bar{p}^2 + 2\bar{p}p - p^2),\ \pm(\bar{p}^2 - 2\bar{p}p + p^2),\ \pm(\bar{p}^2 + 2\bar{p}p + p^2).$$

Which operators besides equivalence generate Euclidian metric?

Solution

$$<p\begin{pmatrix} 1 & 1 \\ -1 & 1 \end{pmatrix}p>,\quad <p\begin{pmatrix} 1 & -1 \\ 1 & 1 \end{pmatrix}p>.$$

Problem

Reduce the tautologies

$$<x\,|\leftrightarrow|\,x>,\quad <x\,|\vee|\,\bar{x}>,\quad <x\,|\rightarrow|\,x>,$$

and the contradictions

$$<x\,|\,\rceil\,|\,x>,\quad <x\,|\wedge|\,\bar{x}>,\quad <x\,|\setminus|\,x>$$

to the canonical form.

Solution

$$<x\,|\leftrightarrow|\,x> \;=\; <\overline{x\,|\,\bar{x}}> \;\equiv 1$$

$$<x\,|\vee|\,\bar{x}> \;=\; <\overline{\bar{\bar{x}}\,|\,1><1\,|\,x}> \;=\; <\overline{\overline{\overline{x\,|\,1}><1\,|\,x}}> \;\equiv 1$$

$$<x\,|\rightarrow|\,x> \;=\; <\overline{x\,|\,1><0\,|\,x}> \;=\; <\overline{x\,|\,1><\overline{\overline{1\,|\,x}}}> \;\equiv 1$$

$$<x\,|\,\rceil\,|\,x> \;=\; <x\,|\,\bar{x}> \;\equiv 0$$

$$<x\,|\wedge|\,\bar{x}> \;=\; <x\,|\,1><1\,|\,\bar{x}> \;=\; <x\,|\,1><\overline{1\,|\,x}> \;\equiv 0$$

$$<x\,|\setminus|\,x> \;=\; <x\,|\,1><0\,|\,x> \;=\; <x\,|\,1><\overline{1\,|\,x}> \;\equiv 0$$

Problem

Construct and prove the tautology: x implies y and y is false, then x is false.

Solution

$$\big((x \to y) \land \bar{y}\big) \to \bar{x}$$

$$\overline{\langle x|\to|y\rangle\langle 1|\bar{y}\rangle\overline{\langle 1|\bar{x}\rangle}} = \overline{\langle x|\to|y\rangle\langle x|1\rangle\langle 0|y\rangle} =$$

$$= \langle\overline{x|\to \ast \backslash}|y\rangle = \langle x|\overline{\hat{0}|y}\rangle \equiv 1$$

Problem

Prove the tautologies

(a) $(x \to \bar{x}) \to \bar{x}$

(b) $x \to (\bar{x} \to y)$

(c) $(x \lor x) \to x$

(d) $x \to (y \lor x)$

Solution

(a) $$\overline{\langle x|\to|\bar{x}\rangle\langle x|1\rangle} = \overline{\langle x|\downarrow|x\rangle\langle x|1\rangle} =$$

$$= \overline{\overline{\langle x|1\rangle\langle 1|x\rangle\langle x|1\rangle}} =$$

$$= \overline{\langle\overline{\overline{x|1}}\rangle\langle x|1\rangle} = \bar{0} \equiv 1$$

(b) $$\overline{\langle x|1\rangle\langle\overline{\bar{x}|\to|y}\rangle} = \overline{\langle x|1\rangle\langle x|\uparrow|y\rangle} =$$

$$= \overline{\langle x|1\rangle\langle x|0\rangle\langle 0|y\rangle} = \bar{0} \equiv 1$$

(c) $$\overline{\langle x|\lor|x\rangle\langle\overline{\overline{x|1}}\rangle} = \overline{\overline{\langle x|0\rangle^2\langle x|0\rangle}} =$$

$$= \overline{\langle\overline{\overline{x|0}}\rangle\langle x|0\rangle} = \bar{0} \equiv 1$$

(d) $$\overline{\langle x|1\rangle\langle\overline{y|\lor|x}\rangle} = \overline{\langle x|1\rangle\langle y|0\rangle\langle 0|x\rangle} = \bar{0} \equiv 1$$

Problem

Prove the tautologies

(a) $(x = y) \rightarrow (x \vee \bar{y})$

(b) $(x = y) \rightarrow (\bar{x} = \bar{y})$

Solution

(a) $$\langle\overline{x \mid y \rangle\langle \overline{x \mid \vee \mid \overline{y}}}\rangle = \langle\overline{x \mid y \rangle\langle x \mid / \mid y}\rangle =$$

$$= \langle\overline{x \mid \hat{0} \mid y}\rangle = \bar{0} \equiv 1$$

(b) $$\langle\overline{x \mid y \rangle\langle \overline{\overline{x} \mid \overline{y}}}\rangle = \langle\overline{x \mid y \rangle\langle \overline{x \mid y}}\rangle = \bar{0} \equiv 1$$

Problem

Prove the tautology $(x \rightarrow y) \vee (y \rightarrow x)$.

Solution

$$\langle\overline{\overline{x \mid \rightarrow \mid y} \rangle\langle \overline{y \mid \rightarrow \mid x}}\rangle = \langle\overline{x \mid \backslash \mid y \rangle\langle y \mid \backslash \mid x}\rangle =$$

$$= \langle\overline{x \mid \backslash \mid y \rangle\langle x \mid / \mid y}\rangle =$$

$$= \langle\overline{x \mid \hat{0} \mid y}\rangle = \bar{0} \equiv 1$$

Problem

Prove that the truth status of y is irrelevant for the tautologies:

(a) $x \rightarrow (y \rightarrow x)$,

(b) $(x \wedge y) \rightarrow x$.

Solution

(a) $x \rightarrow (y \rightarrow x)$

$$\langle\overline{x \mid 1 \rangle\langle y \mid 1 \rangle\langle 0 \mid x}\rangle = \langle\overline{y \mid 1 \rangle \cdot 0} \equiv 1$$

(b) $(x \wedge y) \rightarrow x$

$$\langle\overline{x \mid 1 \rangle\langle 1 \mid y \rangle\langle \overline{x \mid 1}}\rangle = \langle\overline{x \mid 1 \rangle\langle 1 \mid y \rangle\langle x \mid 0}\rangle =$$

$$= \langle\overline{1 \mid y \rangle \cdot 0} \equiv 1$$

Problem

Prove the tautology $x \to (y \to (x \wedge y)$

Solution

$$\langle \overline{x|1\rangle\langle \overline{\overline{y|1\rangle\langle \overline{x|\wedge|y}}}}\rangle = \langle \overline{x|1\rangle\langle 1|y\rangle\langle \overline{x|\wedge|y}}\rangle =$$

$$= \langle \overline{x|\wedge|y\rangle\langle \overline{\overline{x|\wedge|y}}}\rangle \equiv 1$$

Problem

Employ transmultiplication to derive the tautologies

(a) $(x \vee y) \to (y \vee x)$

(b) $(x \to y) \vee (y \to x)$

(c) $(x \wedge y) \to (x \to y)$

(d) $(x \wedge \bar{y}) \to (\overline{x \to y})$

(e) $(x \to \bar{y}) \to (y \to \bar{x})$.

Solution

(a) $$\langle \overline{x|\vee|y\rangle\langle \overline{y|\vee|x}}\rangle = \langle \overline{x|\vee|y\rangle\langle x|\uparrow|y}\rangle =$$

$$= \langle \overline{x|\vee \ast \uparrow|y}\rangle =$$

$$= \langle \overline{x|\hat{0}|y}\rangle \equiv 1$$

(b) $$\langle \overline{\overline{x|\to|y}\rangle\langle \overline{y|\to|x}}\rangle = \langle \overline{x|\backslash|y\rangle\langle x|/|y}\rangle =$$

$$= \langle x|\backslash \ast /|y\rangle =$$

$$= \langle \overline{x|\hat{0}|y}\rangle \equiv 1$$

(c) $$\langle \overline{x|\wedge|y\rangle\langle \overline{\overline{x|\to|y}}}\rangle = \langle \overline{x|\wedge|y\rangle\langle x|\backslash|y}\rangle =$$

$$= \langle x|\wedge \ast \backslash|y\rangle =$$

$$= \langle \overline{x|\hat{0}|y}\rangle \equiv 1$$

(d) $$\langle \overline{x|\wedge|\bar{y}\rangle\langle \overline{\overline{\overline{x|\to|y}}}}\rangle = \langle \overline{x|\backslash|y\rangle\langle x|\to|y}\rangle =$$

$$= \langle x|\backslash \ast \to|y\rangle =$$

$$= \langle \overline{x|\hat{0}|y}\rangle \equiv 1$$

(e)
$$\langle\overline{x|\to|\bar{y}\rangle\langle\overline{y|\to|\bar{x}}}\rangle = \langle\overline{x|\downarrow|y\rangle\langle y|\wedge|x}\rangle =$$
$$= \langle x|\downarrow \ast \wedge|y\rangle =$$
$$= \langle\overline{x|\,\hat{0}\,|y}\rangle \equiv 1$$

Problem

Prove that

(a) $((x \to y) \to x) \to x$

and

(b) $x \to (x \vee y)$

are tautologies of the same form.

Solution

(a)
$$\langle\overline{\overline{x \to y\rangle\langle\overline{x|1}}\rangle\langle\overline{x|1}}\rangle = \langle\overline{\overline{x \to y\rangle\langle x|0}\rangle\langle x|0}\rangle =$$
$$= \langle\overline{x|0\rangle - \langle x \to y\rangle\langle x|0}\rangle =$$
$$= \langle\overline{x|0\rangle(1 - \langle x \to y\rangle)} =$$
$$= \langle\overline{x|0\rangle\langle x \setminus y}\rangle =$$
$$= \langle\overline{x|0\rangle\langle x|1\rangle\langle 0|y}\rangle \equiv 1$$

(b)
$$\langle\overline{x|1\rangle\langle\overline{x|\vee|y}}\rangle = \langle\overline{x|1\rangle\langle x|0\rangle\langle 0|y}\rangle \equiv 1$$

Problem

Prove the legitimacy of exchanging the connectives $\vee$ and $\to$ in the equality

$$x \vee (x \to y) = x \to (x \vee y) .$$

Solution

Both sides of the equality are tautologies of the same form:

$$\langle\overline{\bar{x}|1\rangle\langle x|1\rangle\langle 0|y}\rangle = \langle\overline{x|1\rangle\langle\bar{x}|1\rangle\langle 0|y}\rangle .$$

Problem

Prove the tautology

$$(x \to y) \to [(z \vee x) \to (z \vee y)] .$$

Solution

$$\overline{\langle x|\to|y\rangle\,\overline{\overline{\langle z|\vee|x\rangle\,\overline{\langle z|\vee|y\rangle}}}} =$$

$$= \overline{\langle x|\to|y\rangle\langle z|\vee|x\rangle\langle z|\uparrow|y\rangle} =$$

$$= \overline{\overline{\langle x|1\rangle\langle 0|y\rangle}\,\overline{\langle z|0\rangle\langle 0|x\rangle}\,\langle z|0\rangle\langle 0|y\rangle} =$$

$$= \overline{\langle z|0\rangle\langle 0|y\rangle - \langle x|1\rangle\langle 0|y\rangle\langle z|0\rangle\langle 0|y\rangle -}$$

$$\overline{- \langle z|0\rangle\langle 0|x\rangle\langle z|0\rangle\langle 0|y\rangle +}$$

$$\overline{+ \langle x|1\rangle\langle 0|y\rangle\langle z|0\rangle\langle 0|x\rangle\langle z|0\rangle\langle 0|y\rangle} =$$

$$= \overline{\langle z|0\rangle\langle 0|y\rangle - \langle x|1\rangle\langle 0|y\rangle\langle z|0\rangle -}$$

$$\overline{- \langle z|0\rangle\langle 0|x\rangle\langle 0|y\rangle + 0} =$$

$$= \overline{\langle z|0\rangle\langle 0|y\rangle\,(1 - \langle x|1\rangle - \langle x|0\rangle)} =$$

$$= \overline{\langle z|\uparrow|y\rangle\;0} = \bar{0} \equiv 1 .$$

Problem

Prove the tautology

$$[x \to (y \to z)] \to [(x \to y) \to (x \to z)] .$$

Solution

$$\overline{\overline{\langle x|1\rangle\overline{\langle y|\to|z\rangle}}\;\overline{\overline{\langle x|\to|y\rangle\overline{\langle x|\to|z\rangle}}}} =$$

$$= \overline{\overline{\langle x|1\rangle\langle y|\setminus|z\rangle}\langle x|\to|y\rangle\langle x|\setminus|z\rangle} =$$

$$= \overline{\overline{\langle x|1\rangle\langle y|1\rangle\langle 0|z\rangle}\langle x|\to|y\rangle\langle x|\setminus|z\rangle} =$$

$$= \overline{\overline{\langle y|1\rangle\langle x|\setminus|z\rangle}\langle x|\to|y\rangle\langle x|\setminus|z\rangle} =$$

$$= \overline{\langle x|\to|y\rangle\langle x|\setminus|z\rangle - \langle y|1\rangle\langle x|\setminus|z\rangle\langle x|\to|y\rangle\langle x|\setminus|z\rangle} =$$

$$= \overline{\langle x|\to|y\rangle\langle x|\setminus|z\rangle - \langle y|1\rangle\langle x|\to|y\rangle\langle x|\setminus|z\rangle} =$$

$$= \overline{\langle x|\to|y\rangle\langle x|\setminus|z\rangle(1 - \langle y|1\rangle)} =$$

$$= \overline{\langle x|\to|y\rangle\langle x|1\rangle\langle 0|z\rangle\langle y|0\rangle} =$$

$$= \overline{\langle x|\to|y\rangle\langle x|\setminus|y\rangle\langle 0|z\rangle} =$$

$$= \overline{0\,\langle 0|z\rangle} = \bar{0} \equiv 1 .$$

Problem

Prove the tautology

$$[(x \to y) \to (x \to z)] \to [(y \to x) \to (y \to z)]\,.$$

Solution

$$<\overline{\overline{x|\to|y><\overline{y|\to|z}}><\overline{\overline{y|\to|x><\overline{y|\to|z}}}}> \;=$$

$$= \; <\overline{\overline{x|\to|y><x|\backslash|z}><y|\to|x><y|\backslash|z}> \;=$$

$$= \; <\overline{y|\to|x><y|\backslash|z>-<x|\to|y><x|\backslash|z><y|\to|x><y|\backslash|z}> \;=$$

$$= \; <\overline{y|\to|x><y|\backslash|z>-<x|y><x|\backslash|z><y|\backslash|z}> \;=$$

$$= \; <\overline{y|\backslash|z>-<y|\backslash|x><y|\backslash|z>-<x|y><x|\wedge|y}><z|0> \;=$$

$$= \; <\overline{y|\backslash|z>-<y|\backslash|z><x|0>-<x|\wedge|y}><z|0> \;=$$

$$= \; <\overline{y|\backslash|z>(1-<x|0>)-<x|\wedge|y}><z|0> \;=$$

$$= \; <\overline{y|1><0|z><x|1>-<x|1><1|y}><z|0> \;=\; \overline{0} \equiv 1\,.$$

Problem

Prove that

$$(x \to y) \to ((x \to y) \to y)$$

is the disjunction function.

Solution

$$<\overline{x|\to|y><\overline{\overline{((x \to y) \to y)}}}> \;=$$

$$= \; <\overline{x|\to|y><x|\to|y><0|y}> \;=$$

$$= \; <\overline{x|\to|y><0|y}> \;=$$

$$= \; <\overline{\overline{x|1><0|y}><0|y}> \;=$$

$$= \; <\overline{(1 - <x|1><0|y>)\;<0|y}> \;=$$

$$= \; <\overline{0|y> \;-\; <x|1><0|y>^2} \;=\; <\overline{0|y> \;-\; <x|1><0|y}> \;=$$

$$= \; <\overline{y|0>\;(1 \;-\; <x|1>)} \;=\; <\overline{y|0><\overline{x|1}}> \;=$$

$$= \; <\overline{x|0><0|y}> \;=\; <x|\vee|y>\,.$$

Problem

Prove the equivalence

$$(x \wedge y) \rightarrow z = x \rightarrow (y \rightarrow z) .$$

Solution

$$< \overline{x | \wedge | y >< \overline{z | 1}} > \;=\; < \overline{x | 1 >< \overline{y | \rightarrow | z}} >$$

$$< \overline{x | 1 >< 1 | y >< z | 0} > \;=\; < \overline{x | 1 >< y | 1 >< 0 | z} >$$

Problem

Prove the equivalence

$$(x \rightarrow y) \rightarrow x = x .$$

Solution

$$< \overline{x | \rightarrow | y >< \overline{x | 1}} > \;=\; < x | 1 >$$

$$< x | \rightarrow | y >< x | 0 > \;=\; < \overline{x | 1} >$$

$$< x | 0 > \;-\; < x | 1 >< y | 0 >< x | 0 > \;=\; < x | 0 >$$

$$< x | 0 > \;=\; < x | 0 >$$

Problem

Employing the superassociativity rule, prove the equality

$$x \rightarrow (y \rightarrow z) = \bar{z} \rightarrow (x \downarrow y) .$$

Solution

$$x \rightarrow (y \rightarrow z) \;\Rightarrow\; < \overline{x | 1 >< \overline{y | \rightarrow | z}} > \;=$$

$$= \; < \overline{x | 1 >< y | 1 >< 0 | z} > \;=\; < \overline{z | 0 >< x | 1 >< 1 | y} > \;\Rightarrow$$

$$\Rightarrow \; \bar{z} \rightarrow (x \downarrow y)$$

Problem

Derive the equality

$$(x \rightarrow y) \rightarrow y \;=\; (y \rightarrow x) \rightarrow x$$

Solution

Tautology:

$$x \;=\; x$$

Substitution:

$$<x | 0> <y | 0> \;=\; <<x | 0> <y | 0>$$

Commutation:

$$<y|0><x|0> \ = \ <x|0><y|0>$$

Extraction of complement operator:

$$<y|0><\overline{x|1}> \ = \ <x|0><\overline{y|1}>$$

$$<y|0>-<x|1><y|0> \ = \ <x|0>-<y|1><x|0>$$

Idempotence rule:

$$<y|0>-<x|1><y|0>^2 \ = \ <x|0>-<y|1><x|0>^2$$

$$(1-<x|1><0|y>)<y|0> \ = \ (1-<y|1><0|x>)<x|0>$$

Complementation:

$$<x|\to|y><y|0> \ = \ <y|\to|x><x|0>$$

$$<\overline{x|\to|y><\overline{\overline{y|1}}}> \ = \ <\overline{y|\to|x><\overline{\overline{x|1}}}>$$

Scalarization:

$$(x \to y) \to y \ = \ (y \to x) \to x$$

Problem

Prove the orthomodularity

$$x \to y \to [y \ = \ x \vee (y \wedge \bar{x})] \, .$$

Solution

$$<x|\to|y>\to[<y\to[x \ \vee \ (y \ \wedge \ \bar{x})]><[x \ \vee \ (y \ \wedge \ \bar{x})]\to y>] \ =$$

$$= \ <\overline{x|\to|y><\overline{\overline{y|1><\overline{\overline{x \ \vee \ (y \ \wedge \ \bar{x})}}><\overline{\overline{x \ \vee \ (y \ \wedge \ \bar{x})><\overline{\overline{y|1}}}}}}}> \ =$$

$$= \ <\overline{x|\to|y><\overline{\overline{\overline{\overline{y|1><x|0><y|\to|x}}><\overline{\overline{x|0><y|\to|x}}><y|0}}}> \ =$$

$$= \ <\overline{x|\to|y>\overline{\overline{\bar{0}<\overline{\overline{x|0><y|\to|x}}><y|0}}}> \ =$$

$$= \ <\overline{x|\to|y><\overline{\overline{x|0><y|\to|x}}><y|0}> \ =$$

$$= \ <\overline{x|\to|y><y|0> \ - \ <x|\to|y><y|0><x|0><y|\to|x}> \ =$$

$$= \ <\overline{x|\to|y><y|0> \ - \ <x|y><x|\uparrow|y}> \ =$$

$$= \ <\overline{x|\to|y><y|0> \ - \ <x|\uparrow|y}> \ =$$

$$= \ <\overline{y|0> \ - \ <x|1><0|y> \ - \ <x|\uparrow|y}> \ =$$

$$= \ <\overline{y|0> \ (1 \ - \ <x|1> \ - \ <x|0>)} \ = \ <\overline{y|0> \ 0} \ = \ \bar{0} \equiv 1 \, .$$

Problem

Convolute

$$1 - x - y + xy + x^2y + xy^2 - x^2y^2$$

into a matrix operator product.

Solution

$$1 - x - y + xy + x^2y + xy^2 - x^2y^2 =$$

$$= \overline{x + y - xy - x^2y - xy^2 + x^2y^2} =$$

$$= \overline{x + y - xy - xy(x + y - xy)} =$$

$$= \langle\overline{x|\vee|y\rangle - xy\langle x|\vee|y}\rangle = \langle\overline{x|\vee|y\rangle(1 - xy)} =$$

$$= \langle\overline{x|\vee|y\rangle(1 - \langle x|\wedge|y\rangle)} = \langle\overline{x|\vee|y\rangle\langle\overline{x|\wedge|y}}\rangle =$$

$$= \langle\overline{x|\vee|y\rangle\langle x|\downarrow|y}\rangle = \langle\overline{x|\vee \circledast \downarrow|y}\rangle =$$

$$= \langle\overline{x|\rceil|y}\rangle = \langle x|y\rangle .$$

Problem

Prove the merging of implications and show that for a product of implications of the same antecedent the complement operators can be glued across conjunctions:

$$\langle\overline{x|1\rangle\langle\overline{1|y}}\rangle\langle\overline{x|1\rangle\langle\overline{1|z}}\rangle = \langle\overline{x|1\rangle\langle\overline{y|1\rangle\langle 1|z}}\rangle .$$

Solution

$$\langle x|\rightarrow|y\rangle\langle x|\rightarrow|z\rangle = \langle\overline{x|\backslash|y}\rangle\langle\overline{x|\backslash|z}\rangle =$$

$$= \langle\overline{x|1\rangle\langle\overline{1|y}}\rangle\langle\overline{x|1\rangle\langle\overline{1|z}}\rangle =$$

$$= \langle\overline{x|1\rangle - \langle x|1\rangle\langle 1|y}\rangle\langle\overline{x|1\rangle - \langle x|1\rangle\langle 1|z}\rangle =$$

$$= (1 - \langle x|1\rangle + \langle x|1\rangle\langle 1|y\rangle)\,(1 - \langle x|1\rangle + \langle x|1\rangle\langle 1|z\rangle) =$$

$$= 1 - \langle x|1\rangle + \langle x|1\rangle\langle 1|z\rangle - \langle x|1\rangle + \langle x|1\rangle - \langle x|1\rangle\langle 1|z\rangle +$$

$$+ \langle x|1\rangle\langle 1|y\rangle - \langle x|1\rangle\langle 1|y\rangle + \langle x|1\rangle\langle 1|y\rangle\langle 1|z\rangle =$$

$$= 1 - \langle x|1\rangle + \langle x|1\rangle\langle 1|y\rangle\langle 1|z\rangle =$$

$$= \langle\overline{x|1\rangle - \langle x|1\rangle\langle 1|y\rangle\langle 1|z}\rangle =$$

$$= \langle\overline{x|1\rangle(1 - \langle y|1\rangle\langle 1|z\rangle)} =$$

$$= \langle\overline{x|1\rangle\langle\overline{y|1\rangle\langle 1|z}}\rangle$$

Prove by induction the general case

$$\overline{\alpha\overline{\beta}}\;\overline{\alpha\overline{\gamma}}\;\overline{\alpha\overline{\eta}}\;\ldots = \overline{\alpha\overline{\beta\gamma\eta\ldots}}$$

Problem

Prove the absorption

$$\langle x \mid \uparrow \mid y \rangle \langle x \mid \rightarrow \mid y \rangle = \langle x \mid \uparrow \mid y \rangle$$

by

(a) the transmultiplication rule,

(b) reduction to a fundamental product.

Solution

(a) $$\langle x \mid \uparrow \mid y \rangle \langle x \mid \rightarrow \mid y \rangle = \langle x \mid \uparrow \circledast \rightarrow \mid y \rangle = \langle x \mid \uparrow \mid y \rangle$$

(b) $$\langle x \mid \uparrow \mid y \rangle \langle x \mid \rightarrow \mid y \rangle = \langle x \mid 0 \rangle \langle 0 \mid y \rangle \langle \overline{x \mid \backslash \mid y} \rangle =$$

$$= \langle x \mid 0 \rangle \langle 0 \mid y \rangle (1 - \langle x \mid 1 \rangle \langle 0 \mid y \rangle) =$$

$$= \langle x \mid 0 \rangle \langle 0 \mid y \rangle - \langle x \mid 0 \rangle \langle 0 \mid y \rangle \langle x \mid 1 \rangle =$$

$$= \langle x \mid 0 \rangle \langle 0 \mid y \rangle - 0 = \langle x \mid \uparrow \mid y \rangle$$

Problem

Prove inductively the contraction

$$x \rightarrow (x \rightarrow y) = x \rightarrow y .$$

Solution

Tautology:

$$x = x$$

Substitution:

$$\langle x \mid 1 \rangle \langle 0 \mid y \rangle = \langle x \mid 1 \rangle \langle 0 \mid y \rangle$$

Idempotence rule:

$$\langle x \mid 1 \rangle \langle x \mid 1 \rangle \langle 0 \mid y \rangle = \langle x \mid 1 \rangle \langle 0 \mid y \rangle$$

Extraction of complement operator:

$$\langle x \mid 1 \rangle \langle \overline{x \mid \rightarrow \mid y} \rangle = \langle x \mid 1 \rangle \langle 0 \mid y \rangle$$

Complementation:

$$\overline{\langle x \mid 1 \rangle \langle \overline{x \mid \rightarrow \mid y} \rangle} = \langle \overline{x \mid 1 \rangle \langle 0 \mid y} \rangle$$

Scalarization:

$$x \rightarrow (x \rightarrow y) = x \rightarrow y$$

Problem

Which logic functions represent the following matrix strings

(a) $$\langle x | \overline{\begin{pmatrix}0&0\\0&1\end{pmatrix}\begin{pmatrix}1&1\\0&1\end{pmatrix}\begin{pmatrix}0&1\\1&1\end{pmatrix}\begin{pmatrix}1\\0\end{pmatrix}(0\ \ 1)\begin{pmatrix}1&1\\1&0\end{pmatrix}} | y \rangle\,,$$

(b) $\langle \overline{0 | \vee \wedge \backslash \Omega / \wedge \vee | 0} \rangle$?

Solution

(a) The implication function $\langle x | \rightarrow | y \rangle$ in $\Omega^{\Box}$.

(b) The disjunction function in $\Omega^{\Diamond}$:

$$\langle \overline{0 | \vee \wedge \backslash \Omega / \wedge \vee | 0} \rangle = \langle \overline{0 | \vee \wedge \backslash | p \rangle \langle q | / \wedge \vee | 0} \rangle =$$

$$= \langle \overline{q | / \wedge \vee | 0 \rangle \langle 0 | \vee \wedge \backslash | p} \rangle =$$

$$= \langle \overline{q | / \wedge \vee \uparrow \vee \wedge \backslash | p} \rangle = \langle \overline{q | \uparrow | p} \rangle = \langle p | \vee | q \rangle\,.$$

Problem

Construct virtual transformations for the inhibition function $\langle x | \backslash | y \rangle$ and its complement, the implication function $\langle x | \rightarrow | y \rangle$. Establish the difference between the roles of the virtual term in these two cases.

Solution

For the inhibition function $\langle x | \backslash | y \rangle$:

we *subtract* a contradiction $\langle x | 1 \rangle \langle y | 0 \rangle \langle x | 0 \rangle$:

$$\begin{aligned}\langle x | \backslash | y \rangle &= \langle x | 1 \rangle \langle 0 | y \rangle = \\ &= \langle x | 1 \rangle \langle 0 | y \rangle - \langle x | 1 \rangle \langle y | 0 \rangle \langle x | 0 \rangle = \\ &= (1 - \langle x | 0 \rangle) \langle x | 1 \rangle \langle 0 | y \rangle = \\ &= \langle x | 1 \rangle^2 \langle 0 | y \rangle = \\ &= \langle x | \backslash | y \rangle\end{aligned}$$

For the implication function $\langle x | \rightarrow | y \rangle$:

we *add* a contradiction $\langle x | 1 \rangle \langle y | 0 \rangle \langle x | 0 \rangle$:

$$\begin{aligned}\langle x | \rightarrow | y \rangle &= \langle \overline{x | \backslash | y} \rangle = \\ &= 1 - \langle x | 1 \rangle \langle 0 | y \rangle = \\ &= 1 - \langle x | 1 \rangle \langle 0 | y \rangle + \langle x | 1 \rangle \langle y | 0 \rangle \langle x | 0 \rangle = \\ &= 1 - \langle x | 1 \rangle \langle 0 | y \rangle (1 - \langle x | 0 \rangle) = \\ &= \overline{\langle x | 1 \rangle^2 \langle 0 | y \rangle} = \\ &= \langle x | \rightarrow | y \rangle\end{aligned}$$

Problem

Establish the logic interpretation of the virtual terms

$$\alpha = \langle x \mid 1\rangle\, y \mid 0\rangle \langle x \mid 0\rangle = 0$$

$$\beta = \langle x \mid 1\rangle\, y \mid 1\rangle \langle x \mid 0\rangle = 0$$

Solution

$$\alpha = \overline{\bar{x} \rightarrow (x \rightarrow y)} \quad — \quad \overline{\text{Ex falso quodlibet}}$$

$$\beta = \overline{x \rightarrow (y \rightarrow x)} \quad — \quad \overline{\text{Verum ex quolibet}}$$

Problem

Which octet of the logic operators of positive logic is idempotent?

$$L^2 = L\,.$$

Solution

$$\{I, \wedge, \uparrow, \perp, \top, \dashv, \vdash, \hat{0}\}$$

Problem

Employing the set of filters

$$\{\perp, \top, \dashv, \vdash\}\,,$$

express the discrete logic functions $\langle x \mid L \mid y\rangle$ in terms of the tautology $\hat{1}$ and contradiction $\hat{0}$ operators.

Solution

$$\langle x \mid L \mid y\rangle = \begin{cases} \langle x \mid \dashv L \perp \mid y\rangle & \Rightarrow \quad x = 1,\ y = 1 \\ \langle x \mid \vdash L \perp \mid y\rangle & \Rightarrow \quad x = 0,\ y = 1 \\ \langle x \mid \dashv L \top \mid y\rangle & \Rightarrow \quad x = 1,\ y = 0 \\ \langle x \mid \vdash L \top \mid y\rangle & \Rightarrow \quad x = 0,\ y = 0 \end{cases}$$

Examples

$$L = I$$

$$\langle x \mid I \mid y\rangle = \begin{cases} \langle x \mid \dashv I \perp \mid y\rangle = \langle x \mid \hat{1} \mid y\rangle \\ \langle x \mid \vdash I \perp \mid y\rangle = \langle x \mid \hat{0} \mid y\rangle \\ \langle x \mid \dashv I \top \mid y\rangle = \langle x \mid \hat{0} \mid y\rangle \\ \langle x \mid \vdash I \top \mid y\rangle = \langle x \mid \hat{1} \mid y\rangle \end{cases}$$

$$L = \rceil$$

$$\langle x \mid \rceil \mid y \rangle = \begin{cases} \langle x \mid \dashv \rceil \bot \mid y \rangle = \langle x \mid \hat{0} \mid y \rangle \\ \langle x \mid \vdash \rceil \bot \mid y \rangle = \langle x \mid \hat{1} \mid y \rangle \\ \langle x \mid \dashv \rceil \top \mid y \rangle = \langle x \mid \hat{1} \mid y \rangle \\ \langle x \mid \vdash \rceil \top \mid y \rangle = \langle x \mid \hat{0} \mid y \rangle \end{cases}$$

Problem

Exploring the fact that the effect of the filter (converse filter) upon a logic vector is orthogonal to that of the antifilter (converse antifilter)

$$\bot \mid q\rangle = \mid 1\rangle$$
$$\top \mid q\rangle = \mid 0\rangle$$

$$\langle p \mid \dashv \; = \langle 1 \mid$$
$$\langle p \mid \vdash \; = \langle 0 \mid ,$$

construct annihilating and tautology interactions.

Solution

(a) anti-parallel mode:

$$\langle 0 \mid 1\rangle = \langle p \mid \vdash \bot \mid q\rangle = \langle p \mid \hat{0} \mid q\rangle \equiv 0$$
$$\langle 1 \mid 0\rangle = \langle p \mid \dashv \top \mid q\rangle = \langle p \mid \hat{0} \mid q\rangle \equiv 0$$

(b) parallel mode:

$$\langle 0 \mid 0\rangle = \langle p \mid \vdash \top \mid q\rangle = \langle p \mid \hat{1} \mid q\rangle \equiv 1$$
$$\langle 1 \mid 1\rangle = \langle p \mid \dashv \bot \mid q\rangle = \langle p \mid \hat{1} \mid q\rangle \equiv 1$$

Which operations relate matrix multipliers in the parallel and the anti-parallel mode?

Solution

(a) converse transposition:

$$\overline{\vdash} = \bot$$

$$\overline{\dashv} = \top$$

(b) transposition:

$$\tilde{\vdash} = \top$$
$$\tilde{\dashv} = \bot$$

With respect to which operations are annihilating and tautology interactions

(1) unchanged,

(2) functionally invariant ?

Solution

(1) $\overline{\vdash \bot} = \overline{\bot}\;\overline{\vdash} = \vdash \bot$

$\overline{\dashv \top} = \overline{\top}\;\overline{\dashv} = \dashv \top$

$\widetilde{\vdash \top} = \tilde{\top}\;\tilde{\vdash} = \vdash \top$

$\widetilde{\dashv \bot} = \tilde{\bot}\;\tilde{\dashv} = \dashv \bot$

(2) $\widetilde{\vdash \bot} = \dashv \top = \hat{0}$

$\overline{\vdash \top} = \dashv \bot = \hat{1}$

Problem

The logic norm of the universum operator Ω is tautology. Consider the universum operator in the case of 2, 3, 4 and 5 logic vectors. Derive the corresponding duality relations and show that for $n > 3$ the complement operator in its standard form is not applicable. Construct the complement operator for the case of 4 and 5 variable logic vectors.

Solution

n = 2

$$\sum_{i,j}^{2} p_i \otimes q_j \equiv 1$$

$$\bar{p}\bar{q} + \bar{p}q + p\bar{q} + pq = 1$$

$$pq = 1 - \bar{p}\bar{q} - \bar{p}q - p\bar{q} = \overline{\bar{p}\bar{q} + \bar{p}q + p\bar{q}}$$

$$< p | \wedge | q > \; = \; < \overline{\bar{p} | \vee | \bar{q}} >$$

n = 3

$$\sum_{i,j,k}^{2} p_i \otimes q_j \otimes r_k \equiv 1$$

$$(\bar p \bar q + \bar p q + p \bar q + p q) r + (\bar p \bar q + \bar p q + p \bar q + p q) \bar r = 1$$

$$\bar p \bar q r + \bar p q r + p \bar q r + p q r + \bar p \bar q \bar r + \bar p q \bar r + p \bar q \bar r + p q \bar r = 1$$

$$\bar p \bar q r + \bar p q r + p \bar q r + p q (r + \bar r) = \overline{\bar p \bar q \bar r + \bar p q \bar r + p \bar q \bar r}$$

$$< p \begin{pmatrix} r & r \\ r & 1 \end{pmatrix} q > \; = \; \overline{< \bar p \begin{pmatrix} 0 & \bar r \\ \bar r & \bar r \end{pmatrix} \bar q >}$$

Similarly

$$< p \begin{pmatrix} q & q \\ q & 1 \end{pmatrix} r > \; = \; \overline{< \bar p \begin{pmatrix} 0 & \bar q \\ \bar q & \bar q \end{pmatrix} \bar r >}$$

$$< q \begin{pmatrix} p & p \\ p & 1 \end{pmatrix} r > \; = \; \overline{< \bar q \begin{pmatrix} 0 & \bar p \\ \bar p & \bar p \end{pmatrix} \bar r >}$$

For n = 3, still just as for n = 2, the complement operator is employed in standard form:

$$\bar L = \hat 1 - L \, .$$

For n > 3 this is no longer possible: the standard relation breaks down.

n = 4

$$\sum_{i,j,k,m}^{2} p_i \otimes q_j \otimes r_k \otimes s_m \equiv 1$$

$$(\bar p \bar q r + \bar p q r + p \bar q r + p q r) s + (\bar p \bar q \bar r + \bar p q \bar r + p \bar q \bar r + p q \bar r) \bar s = 1$$

$$\bar p \bar q r s + \bar p q r s + p \bar q r s + p q (r s + \bar r \bar s) = \overline{\bar p \bar q \bar r s + \bar p q \bar r \bar s + p \bar q \bar r \bar s}$$

$$< p \begin{pmatrix} rs & rs \\ rs & rs + \bar r \bar s \end{pmatrix} q > \; = \; \overline{< \bar p \begin{pmatrix} 0 & \bar r \bar s \\ \bar r \bar s & \bar r \bar s \end{pmatrix} \bar q >}$$

The complement operator here takes the form of subtraction from the matrix

$$(r s + \bar r \bar s) \, \hat 1$$

$\boxed{n = 5}$

$$\sum_{i,j,k,m,n}^{2} p_i \otimes q_j \otimes r_k \otimes s_m \otimes t_n \equiv 1$$

The complement operator takes the form of subtraction from the matrix

$$(r\,s\,t + \bar{r}\,\bar{s}\,\bar{t})\,\hat{1}$$

Solve the problem for $n \to \infty$.

Problem

Compare the conventional and matrix logic forms of the logic formula

$$(x \to y) \to z$$

expressed in the minimal basis set $\{ \downarrow \}$.

Solution

<u>conventional logic</u>:

$$(x \to y) \to z \;=\; (x \downarrow \bar{y}) \downarrow \bar{z} \;=\; (x \downarrow (y \downarrow y)) \downarrow (z \downarrow z)$$

Note: 4 connectives appear in the resulting expression, two more than in the initial expression.

<u>matrix logic</u>:

$$\begin{aligned} < \overline{\overline{x \mid 1 > < 0 \mid y} > < \overline{1} \mid z} > &= < \overline{\overline{x \mid 1 > < 0 \mid y} > < 0 \mid z} > = \\ &= \overline{< < x \mid 1 > \mid \downarrow \mid < 0 \mid y >> < 0 \mid z >} = \\ &= <<< x \mid 1 > \mid \downarrow \mid < 0 \mid y >> \mid \downarrow \mid < 0 \mid z >> \end{aligned}$$

Note: 2 connective operators appear in the resulting expression, the same number as in the initial expression.

Problem

Solve the following expression by

(a) vectorization

and

(b) transmultiplication:

$$\alpha = \overline{<x|\vee|y><x|\downarrow|y>} \; .$$

Solution

(a)
$$\alpha = <<x|\vee|y>|\downarrow|<x|\downarrow|y>> =$$

$$= \left(<\overline{x|\vee|y}>,\; <x|\vee|y>\right)\begin{pmatrix}1 & 1\\ 1 & 0\end{pmatrix}\begin{pmatrix}<\overline{x|\downarrow|y}>\\ <x|\downarrow|y>\end{pmatrix} =$$

$$= \left(1,\; <x|\uparrow|y>\right)\begin{pmatrix}<x|\wedge|y>\\ <x|\downarrow|y>\end{pmatrix} =$$

$$= <x|\wedge|y> + <x|\uparrow|y><x|\downarrow|y> =$$

$$= <x|\wedge + \uparrow|y> \; = \; <x|y>$$

(b)
$$\alpha = \overline{<x|\vee|y><x|\downarrow|y>} =$$

$$= \overline{<x|\vee \circledast \downarrow|y>} =$$

$$= \overline{<x|\rceil|y>} \; = \; <x|y>$$

Problem

According to the prime number axiom $x^{\alpha} = x$, obtain a contradiction law for the logic exponent $\alpha = 5$.

Solution

$x^5 = x$

$x - x^5 = 0$

$x(1 - x^4) = 0$

$x_1 = 0, \qquad x_2 = 1, \qquad x_3 = -1, \qquad x_4 = i$

$x_1 = 0$	—	false
$x_2 = 1$	—	true
$x_3 = -1$	—	antitrue
$x_4 = i$	—	imaginary

Problem

Among the 81 logic operators find 40 non-singular operators whose inverses are defined in the integer set $E_3 = \{-1, 0, 1\}$. Write down all 27 products LL^{-1} generating the unit operator,

$$LL^{-1} = I\,.$$

Solution

$$\begin{pmatrix}1&0\\0&1\end{pmatrix}^2, \begin{pmatrix}0&1\\1&0\end{pmatrix}^2, \begin{pmatrix}-1&0\\0&1\end{pmatrix}^2, \begin{pmatrix}1&0\\0&-1\end{pmatrix}^2, \begin{pmatrix}-1&0\\0&-1\end{pmatrix}^2, \begin{pmatrix}0&-1\\-1&0\end{pmatrix}^2, \begin{pmatrix}-1&0\\1&1\end{pmatrix}^2,$$

$$\begin{pmatrix}1&0\\1&-1\end{pmatrix}^2, \begin{pmatrix}1&0\\-1&-1\end{pmatrix}^2, \begin{pmatrix}-1&0\\-1&1\end{pmatrix}^2, \begin{pmatrix}1&1\\0&-1\end{pmatrix}^2, \begin{pmatrix}-1&1\\0&1\end{pmatrix}^2, \begin{pmatrix}-1&-1\\0&1\end{pmatrix}^2, \begin{pmatrix}1&-1\\0&-1\end{pmatrix}^2$$

$$\begin{pmatrix}0&-1\\1&0\end{pmatrix}\begin{pmatrix}0&1\\-1&0\end{pmatrix}, \begin{pmatrix}0&1\\-1&1\end{pmatrix}\begin{pmatrix}1&-1\\1&0\end{pmatrix}, \begin{pmatrix}0&-1\\1&1\end{pmatrix}\begin{pmatrix}1&1\\-1&0\end{pmatrix}, \begin{pmatrix}0&1\\-1&-1\end{pmatrix}\begin{pmatrix}-1&-1\\1&0\end{pmatrix},$$

$$\begin{pmatrix}0&-1\\1&-1\end{pmatrix}\begin{pmatrix}-1&1\\-1&0\end{pmatrix}, \begin{pmatrix}0&-1\\-1&1\end{pmatrix}\begin{pmatrix}-1&-1\\-1&0\end{pmatrix}, \begin{pmatrix}0&-1\\-1&-1\end{pmatrix}\begin{pmatrix}1&-1\\-1&0\end{pmatrix}, \begin{pmatrix}-1&0\\1&-1\end{pmatrix}\begin{pmatrix}-1&0\\-1&-1\end{pmatrix},$$

$$\begin{pmatrix}-1&1\\0&-1\end{pmatrix}\begin{pmatrix}-1&-1\\0&-1\end{pmatrix}, \begin{pmatrix}0&1\\1&1\end{pmatrix}\begin{pmatrix}-1&1\\1&0\end{pmatrix}, \begin{pmatrix}1&1\\1&0\end{pmatrix}\begin{pmatrix}0&1\\1&-1\end{pmatrix}, \begin{pmatrix}1&1\\0&1\end{pmatrix}\begin{pmatrix}1&-1\\0&1\end{pmatrix},$$

$$\begin{pmatrix}1&0\\1&1\end{pmatrix}\begin{pmatrix}1&0\\-1&1\end{pmatrix}$$

Problem

Employ the results of the preceding problem to compute 38 products which generate the negation operator.

Hint

$$(\urcorner L)\, L^{-1} = L\,(L^{-1}\urcorner) = \urcorner\,.$$

Problem

How many forbidden and how many orthogonal interactions are there in the multiplication subtable of the tautology series?

Solution

Of the 256 possible products 224 products are forbidden and 32 are orthogonal.

Problem

Prove that the tautology operator and the non-integer inverses of the tautology series can be employed as the basis set of connectives.

Solution

The operators are sufficient to obtain the operators of the matrix logic base:

$$\frac{1}{2}\begin{pmatrix}-1 & 1\\ 1 & 1\end{pmatrix}\begin{pmatrix}1 & 1\\ 1 & 1\end{pmatrix}\frac{1}{2}\begin{pmatrix}-1 & 1\\ 1 & 1\end{pmatrix} = \wedge$$

$$\frac{1}{2}\begin{pmatrix}1 & -1\\ 1 & 1\end{pmatrix}\begin{pmatrix}1 & 1\\ 1 & 1\end{pmatrix}\frac{1}{2}\begin{pmatrix}1 & -1\\ 1 & 1\end{pmatrix} = \backslash$$

$$\frac{1}{2}\begin{pmatrix}1 & 1\\ -1 & 1\end{pmatrix}\begin{pmatrix}1 & 1\\ 1 & 1\end{pmatrix}\frac{1}{2}\begin{pmatrix}1 & 1\\ -1 & 1\end{pmatrix} = /$$

$$\frac{1}{2}\begin{pmatrix}1 & 1\\ 1 & -1\end{pmatrix}\begin{pmatrix}1 & 1\\ 1 & 1\end{pmatrix}\frac{1}{2}\begin{pmatrix}1 & 1\\ 1 & -1\end{pmatrix} = \uparrow$$

Problem

In positive logic the equivalence function represents the Kronecker function:

$$\langle x|\leftrightarrow|y\rangle = \delta_{xy} = \begin{cases} 0 & \text{for } x \neq y \\ 1 & \text{for } x = y \end{cases}.$$

Which logic function $L^{-}(x,y)$ behaves like the Kronecker function for the positive-negative logic set $\{-1,1\}$?

Solution

$$L^{-}(x,y) = \langle x\begin{pmatrix}-1 & 1\\ 1 & -1\end{pmatrix}y\rangle = \langle\overline{x|y}\rangle - \langle x|y\rangle =$$

$$= 1 - \langle x|y\rangle - \langle x|y\rangle = 1 - 2\langle x|y\rangle =$$

$$= \overline{2\langle x|y\rangle} = \begin{cases} 1 & \text{for } x \neq y \\ -1 & \text{for } x = y \end{cases}$$

Prove that the function $L^{-}(x,y)$ equals the determinant of the complement of the universum operator of discrete logic

$$L^{-}(x,y) = \det | \overline{\Omega^{\square}} | .$$

Solution

$$\begin{aligned}
\det \overline{\Omega^{\square}} &= \det \overline{|x><y|} = \\
&= \overline{\bar{x}\,\bar{y}}\;\overline{x\,y} - \overline{\bar{x}\,y}\;\overline{x\,\bar{y}} = \\
&= 1 - \bar{x}\,\bar{y} - x\,y + \bar{x}\,\bar{y}\,x\,y - 1 + \bar{x}\,y + x\,\bar{y} - \bar{x}\,y\,x\,\bar{y} = \\
&= \bar{x}\,y + x\,\bar{y} - \bar{x}\,\bar{y} - x\,y = \\
&= <\overline{x\,|\,y}> - <x\,|\,y> = \\
&= \overline{2<x\,|\,y>}
\end{aligned}$$

Problem

Find the additive representation for the equivalence function

$$<x\,|\,y> = <x\,|\rightarrow|\,y>\,<x\,|\leftarrow|\,y>$$

Solution

$$<x\,|\,y> = \tfrac{1}{2}\,(<x\,|\rightarrow|\,y> + <x\,|\rightarrow^{-1}|\,y>)$$

Problem

In the undecided universe, $p = q = \frac{1}{2}$, how many of the 81 logic operators are equivalent to the empty operator $\hat{0}$?

Solution

19 logic operators:

$$\begin{pmatrix}0&0\\0&0\end{pmatrix}, \begin{pmatrix}-1&0\\0&1\end{pmatrix}, \begin{pmatrix}1&0\\0&-1\end{pmatrix}, \begin{pmatrix}0&-1\\1&0\end{pmatrix}, \begin{pmatrix}0&1\\-1&0\end{pmatrix},$$

$$\begin{pmatrix}0&0\\-1&1\end{pmatrix}, \begin{pmatrix}0&0\\1&-1\end{pmatrix}, \begin{pmatrix}-1&1\\0&0\end{pmatrix}, \begin{pmatrix}1&-1\\0&0\end{pmatrix},$$

$$\begin{pmatrix}0&-1\\0&1\end{pmatrix}, \begin{pmatrix}0&1\\0&-1\end{pmatrix}, \begin{pmatrix}-1&0\\1&0\end{pmatrix}, \begin{pmatrix}1&0\\-1&0\end{pmatrix},$$

$$\begin{pmatrix} -1 & 1 \\ 1 & -1 \end{pmatrix}, \begin{pmatrix} 1 & -1 \\ -1 & 1 \end{pmatrix}, \begin{pmatrix} 1 & 1 \\ -1 & -1 \end{pmatrix}, \begin{pmatrix} -1 & -1 \\ 1 & 1 \end{pmatrix}, \begin{pmatrix} 1 & -1 \\ 1 & -1 \end{pmatrix}, \begin{pmatrix} -1 & 1 \\ -1 & 1 \end{pmatrix}.$$

Prove that in the undecided universe the value of the logic functions depends on the norm of an operator only.

Solution

Operator norm	Number of functions	Truth-value
-4	1	-1
-3	4	$-\frac{3}{2}$
-2	10	$-\frac{1}{2}$
-1	16	$-\frac{1}{4}$
0	19	0
1	16	$\frac{1}{4}$
2	10	$\frac{1}{2}$
3	4	$\frac{3}{2}$
4	1	1

Problem

In self-referential logic paradoxes (the liar paradox, the Russell paradox, etc.) the classical rule of inference periodically enforces two mutually exclusive propositions. Construct a periodic autoproduct of autoproducts which can accommodate these phenomena.

Solution

$$\underbrace{\overbrace{<1|\dashv|0>}^{\times}}_{\tau_1}\ \underbrace{\overbrace{<0|\vdash|1>}^{\times}}_{\tau_2} \quad (\text{outer product } \times) = \begin{cases} \dashv & \text{for } \tau_1 \\ \vdash & \text{for } \tau_2 \end{cases}$$

Problem

Evaluate the multiplication table of matrix logic operators employing the rule for Boolean addition, $1 + 1 = 1$.

Compare it with the physical multiplication table and the addition mod 2 multiplication table.

Prove the logic addition duality laws for logic operators

$$L_1 \circledast L_2 = \overline{\overline{\overline{L_1} + \overline{L_2}}}$$

$$L_1 + L_2 = \overline{\overline{\overline{L_1} \circledast \overline{L_2}}}$$

where $\circledast$ is transmultiplication and + logic addition.

Solution

×	↔	⌉	∧	∨	→	←	↓	↑	\	/	⊥	⊤	⊣	⊢	**1**	**0**
↔	↔	⌉	∧	∨	→	←	↓	↑	\	/	⊥	⊤	⊣	⊢	**1**	**0**
⌉	⌉	↔	/	→	∨	↓	←	\	↑	∧	⊤	⊥	⊣	⊢	**1**	**0**
∧	∧	\	∧	⊥	∧	⊥	\	**0**	\	**0**	⊥	**0**	∧	\	⊥	**0**
∨	∨	←	⊣	**1**	∨	**1**	←	\	⊢	∧	**1**	⊥	⊣	⊢	**1**	**0**
→	→	↓	⊣	**1**	→	**1**	↓	↑	⊢	/	**1**	⊤	⊣	⊢	**1**	**0**
←	←	∨	∧	∨	**1**	←	**1**	⊢	\	⊣	⊥	**1**	⊣	⊢	**1**	**0**
↓	↓	→	/	→	**1**	↓	**1**	⊢	↑	⊣	⊤	**1**	⊣	⊢	**1**	**0**
↑	↑	/	**0**	/	⊤	↑	⊤	↑	**0**	/	**0**	⊤	/	↑	⊤	**0**
\	\	∧	**0**	∧	⊥	\	⊥	\	**0**	∧	**0**	⊥	∧	\	⊥	**0**
/	/	↑	/	⊤	/	⊤	↑	**0**	↑	**0**	⊤	**0**	/	↑	⊤	**0**
⊥	⊥	⊥	∧	⊥	⊥	⊥	⊥	\	\	∧	⊥	⊥	∧	\	⊥	**0**
⊤	⊤	⊤	/	⊤	⊤	⊤	⊤	↑	↑	/	⊤	⊤	/	↑	⊤	**0**
⊣	⊣	⊢	⊣	**1**	⊣	**1**	⊢	**0**	⊢	**0**	**1**	**0**	⊣	⊢	**1**	**0**
⊢	⊢	⊣	**0**	⊣	**1**	⊢	**1**	⊢	**0**	⊣	**0**	**1**	⊣	⊢	**1**	**0**
1	**1**	**1**	⊣	**1**	**1**	**1**	**1**	⊢	⊢	⊣	**1**	**1**	⊣	⊢	**1**	**0**
0	**0**	**0**	**0**	**0**	**0**	**0**	**0**	**0**	**0**	**0**	**0**	**0**	**0**	**0**	**0**	**0**

Logic addition, $1 + 1 = 1$, multiplication table

To prove the duality laws it is sufficient to establish the following relations for numbers:

$$\begin{cases} 1 \cdot 1 = \overline{\bar{1} + \bar{1}} = \bar{0} = 1 \\ 0 \cdot 1 = \overline{\bar{0} + \bar{1}} = \bar{1} = 0 \\ 0 \cdot 0 = \overline{\bar{0} + \bar{0}} = \overline{1 + 1} = \bar{1} = 0 \end{cases}$$

$$\begin{cases} 1 + 1 = \overline{\bar{1} \cdot \bar{1}} = \bar{0} = 1 \\ 0 + 1 = \overline{\bar{0} \cdot \bar{1}} = \bar{0} = 1 \\ 0 + 0 = \overline{\bar{0} \cdot \bar{0}} = \bar{1} = 0 \end{cases}$$

Problem

Solve the operator string problems and establish the equality of the following matrix logic expressions

$$\langle p | \vee | 0 \rangle \langle 1 | \rightarrow | q \rangle \quad \text{and} \quad \langle p | \wedge | 1 \rangle \langle 1 | \rightarrow | q \rangle$$

Solution

$$\langle p | \vee | 0 \rangle \langle 1 | \rightarrow | q \rangle = \langle p | \vee / \rightarrow | q \rangle = \langle p | \wedge | q \rangle$$

$$\langle p | \wedge | 1 \rangle \langle 1 | \rightarrow | q \rangle = \langle p | \wedge \wedge \rightarrow | q \rangle = \langle p | \wedge | q \rangle$$

Thus

$$\langle p | \vee | 0 \rangle \langle 1 | \rightarrow | q \rangle = \langle p | \wedge | 1 \rangle \langle 1 | \rightarrow | q \rangle$$

Employing the operator formulation, verify that

(a) the disjunction function $\vee(p,q)$ represents $\max(p,q)$, and

(b) the conjunction function $\wedge(p,q)$ is $\min(p,q)$.

Solution

(a) $$\langle p | \vee | 1 \rangle = \begin{pmatrix} \bar{p} & p \end{pmatrix} \begin{pmatrix} 1 \\ 1 \end{pmatrix} = \bar{p} + p = 1$$

$$\langle p | \vee | 0 \rangle = \begin{pmatrix} \bar{p} & p \end{pmatrix} \begin{pmatrix} 0 \\ 1 \end{pmatrix} = p$$

(b) $$\langle p | \wedge | 1 \rangle = \begin{pmatrix} \bar{p} & p \end{pmatrix} \begin{pmatrix} 0 \\ 1 \end{pmatrix} = p$$

$$\langle p | \wedge | 0 \rangle = \begin{pmatrix} \bar{p} & p \end{pmatrix} \begin{pmatrix} 0 \\ 0 \end{pmatrix} = 0$$

Employ these results to solve the previous problem without resorting to computation of operator strings.

Solution

Since $<p \mid \vee \mid 0> \; = \; <p \mid \wedge \mid 1>$ we establish directly

$$<p \mid \vee \mid 0><1 \mid \rightarrow \mid q> \; = \; <p \mid \wedge \mid 1><1 \mid \rightarrow \mid q> .$$

Problem

In which ket spaces of virtual logic vectors will the operators V, →, ←, ↓ act as equivalence operators?

Solution

$$< p \mid \vee \mid \begin{matrix} q - \bar{q} \\ \bar{q} \end{matrix} > \; = \; < p \mid q > \; = \; < p \mid \vee \, \vee^{-1} \mid q >$$

$$< p \mid \rightarrow \mid \begin{matrix} \bar{q} - q \\ q \end{matrix} > \; = \; < p \mid q > \; = \; < p \mid \rightarrow \, \rightarrow^{-1} \mid q >$$

$$< p \mid \leftarrow \mid \begin{matrix} \bar{q} \\ q - \bar{q} \end{matrix} > \; = \; < p \mid q > \; = \; < p \mid \leftarrow \, \leftarrow^{-1} \mid q >$$

$$< p \mid \downarrow \mid \begin{matrix} q \\ \bar{q} - q \end{matrix} > \; = \; < p \mid q > \; = \; < p \mid \downarrow \, \downarrow^{-1} \mid q >$$

Problem

What is the logical meaning of the derivatives of the equivalence and inequivalence functions?

$$< p \mid q >'_p \; = \; 2q - 1$$

$$< \overline{p \mid q} >'_p \; = \; 1 - 2q$$

Solution

The derivatives represent the difference between the verum and falsum components of the logic vectors

$$q - \bar{q} = 2q - 1 = \; < p \mid q >'_p$$

$$\bar{q} - q = 1 - 2q = \; < \overline{p \mid q} >'_p$$

Prove that the derivative of equivalence with respect to verum equals the derivative of inequivalence with respect to falsum and vice versa.

Solution

$$< p \mid q >'_{\bar{p}} = (\bar{p}\,\bar{q} + p\,q)'_{\bar{p}} = (\bar{p}\,\bar{q} + \bar{\bar{p}}\,q)'_{\bar{p}} = \bar{q} - q = \overline{< p \mid q >}'_{p}$$

$$\overline{< p \mid q >}'_{\bar{p}} = (\bar{p}\,q + p\,\bar{q})'_{\bar{p}} = (\bar{p}\,q + \bar{\bar{p}}\,\bar{q})'_{\bar{p}} = q - \bar{q} = < p \mid q >'_{p}$$

Problem

At which values of the verum and falsum components is the square of the Euclidian norm of the logic vector

$$r^2 = \bar{p}^2 + p^2$$

at a minimum?

Solution

Differentiating r^2 with respect to p and setting the result to 0:

$$\left(\bar{p}^2 + p^2\right)'_p = \left[(1 - p)^2 + p^2\right]'_p =$$

$$= \left(1 - 2p + p^2 + p^2\right)'_p =$$

$$= -2 + 4p = 0$$

$$p = \tfrac{1}{2} \qquad \bar{p} = \tfrac{1}{2}$$

The minimum occurs in the undecided state $\bar{p} = p = \frac{1}{2}$.

Problem

Express the logic bracket products

$$< p \begin{pmatrix} 1 & n \\ 0 & 1 \end{pmatrix} q > \quad \text{and} \quad < p \begin{pmatrix} 1 & 0 \\ n & 1 \end{pmatrix} q >$$

as operator strings and show that the bracket operators can be employed as counters of implications and converse implications in the string.

Solution

$$<p\begin{pmatrix}1 & n\\ 0 & 1\end{pmatrix}q> \;=\; <p\,|\underbrace{\rightarrow\;\rightarrow\;\dots\;\rightarrow}_{\text{n times}}|\,q>$$

$$<p\begin{pmatrix}1 & 0\\ n & 1\end{pmatrix}q> \;=\; <p\,|\underbrace{\leftarrow\;\leftarrow\;\dots\;\leftarrow}_{\text{n times}}|\,q>$$

Problem

From the equality

$$<p\,|\,L\,|\,q> \;=\; <p\,|\,x><y\,|\,q>$$

obtain the universum operators Ω, employing

(a) commutativity of the scalar product,

(b) associativity of number multiplication,

and examine the difference between the operators.

Solution

(a) $<p\,|\,x><y\,|\,q> = <\tilde{p}\,|\,x><\tilde{y}\,|\,q> = <x\,|\,p><q\,|\,y> = <x\,|\,\Omega\,|\,y>$

(b) $<p\,|\,x><y\,|\,q> = <y\,|\,q><p\,|\,x> = <x\,\widetilde{|\,\Omega\,|}\,y>$

Problem

Prove the relation between the tautology operator $\hat{1}$ and the universum operator Ω

$$<p\,|\,\hat{1}\,|\,q> \;=\; <\mathbf{1}\,|\,\Omega\,|\,\mathbf{1}>.$$

Solution

$$\begin{aligned}<p\,|\,\hat{1}\,|\,q> &= <p\,|\uparrow|\,q> + <p\setminus q> + <p\,/\,q> + <p\wedge q> \;=\\ &= <0\,|\,\Omega\,|\,0> + <1\,|\,\Omega\,|\,0> + <0\,|\,\Omega\,|\,1> + <1\,|\,\Omega\,|\,1> \;=\\ &= (<0\,|\,\Omega + <1\,|\,\Omega)\,|\,0> + (<0\,|\,\Omega + <1\,|\,\Omega)\,|\,1> \;=\\ &= (<0\,|\,\Omega + <1\,|\,\Omega)\,(|\,0> + |\,1>) \;=\\ &= (<0\,|\,\Omega + <1\,|\,\Omega)\begin{pmatrix}1\\1\end{pmatrix} =\\ &= (<0\,| + <1\,|\,)\,\Omega\begin{pmatrix}1\\1\end{pmatrix} =\end{aligned}$$

$$= (\,1\;\;1\,)\,\Omega \begin{pmatrix} 1 \\ 1 \end{pmatrix} =$$

$$= <\mathbf{1} \,|\, \Omega \,|\, \mathbf{1}> .$$

Simplify the above derivation by employing the outer product representation for the tautology operator.

Solution

$$<p \,|\, \hat{1} \,|\, q> \;=\; <p \,|\, \mathbf{1}><\mathbf{1} \,|\, q> \;=\; <\mathbf{1} \,|\, p><q \,|\, \mathbf{1}> \;=\; <\mathbf{1} \,|\, \Omega \,|\, \mathbf{1}>$$

Problem

Prove that $\mathrm{Tr}\,\Omega$ is invariant to the orthogonal transformation of the universum operator

$$\mathrm{Tr}\,\Omega = \mathrm{Tr}\,(\theta\Omega) .$$

Solution

$$\mathrm{Tr}\,\Omega = \mathrm{Tr} \,|\, p><q \,| = \overline{\mathrm{Tr} \,|\, \bar{p}><q \,|} = \overline{\overline{\mathrm{Tr} \,|\, \bar{p}><\bar{q} \,|}} =$$

$$= \mathrm{Tr} \,|\, \bar{p}><\bar{q} \,| = \mathrm{Tr}\,\theta\,(|\, p><q \,|) = \mathrm{Tr}\,(\theta\,\Omega) .$$

Problem

Prove that the trace of the tensor product of two universum operators is the product of the traces of these operators:

$$\mathrm{Tr}\,(\Omega_{pq} \otimes \Omega_{rs}) = \mathrm{Tr}\,\Omega_{pq}\;\mathrm{Tr}\,\Omega_{rs} .$$

Solution

$$\mathrm{Tr}\,(\Omega_{pq} \otimes \Omega_{rs}) = \mathrm{Tr} \begin{pmatrix} \bar{p}\bar{q}\bar{r}\bar{s} & \bar{p}\bar{q}\bar{r}s & \bar{p}q\bar{r}\bar{s} & \bar{p}q\bar{r}s \\ \bar{p}\bar{q}r\bar{s} & \bar{p}\bar{q}rs & \bar{p}qr\bar{s} & \bar{p}qrs \\ p\bar{q}\bar{r}\bar{s} & p\bar{q}\bar{r}s & pq\bar{r}\bar{s} & pq\bar{r}s \\ p\bar{q}r\bar{s} & p\bar{q}rs & pqr\bar{s} & pqrs \end{pmatrix} =$$

$$= \bar{p}\bar{q}\bar{r}\bar{s} + \bar{p}\bar{q}rs + pq\bar{r}\bar{s} + pqrs =$$

$$= \bar{p}\bar{q}(\bar{r}\bar{s} + rs) + pq(\bar{r}\bar{s} + rs) =$$

$$= (\bar{p}\bar{q} + pq)(\bar{r}\bar{s} + rs) =$$

$$= <p \,|\, q><r \,|\, s> =$$

$$= \mathrm{Tr}\,\Omega_{pq}\;\mathrm{Tr}\,\Omega_{rs} .$$

Compare this with the trace of the matrix product $\mathrm{Tr}\,(\Omega_{pq}\,\Omega_{rs})$:

$$\begin{aligned}\mathrm{Tr}\,(\Omega_{pq}\;\Omega_{rs}) &= \mathrm{Tr}\,|\,p\rangle\langle q\,|\,r\rangle\langle s\,| = \\ &= \langle q\,|\,r\rangle\,\mathrm{Tr}\,|\,p\rangle\langle s\,| = \\ &= \langle q\,|\,r\rangle\langle p\,|\,s\rangle\,.\end{aligned}$$

Problem

Which operation describes time reversal in matrix logic? How does time reversal alter logic operators?

Solution

$$L(-\tau) = \tilde{L}(\tau)$$

$L = \tilde{L}$	$L \neq \tilde{L}$
$I(-\tau) = I(\tau)$	$\rightarrow(-\tau) = \leftarrow(\tau)$
$\urcorner(-\tau) = \urcorner(\tau)$	$\leftarrow(-\tau) = \rightarrow(\tau)$
$\hat{1}(-\tau) = \hat{1}(\tau)$	$\backslash(-\tau) = /(\tau)$
$\hat{0}(-\tau) = \hat{0}(\tau)$	$/(-\tau) = \backslash(\tau)$
$\wedge(-\tau) = \wedge(\tau)$	$\perp(-\tau) = \dashv(\tau)$
$\vee(-\tau) = \vee(\tau)$	$\mathrm{T}(-\tau) = \vdash(\tau)$
$\downarrow(-\tau) = \downarrow(\tau)$	$\dashv(-\tau) = \perp(\tau)$
$\uparrow(-\tau) = \uparrow(\tau)$	$\vdash(-\tau) = \mathrm{T}(\tau)$

Problem

Prove that the negation and complement operators commute.

Solution

$$[\neg, {}^{-}] = (\neg{}^{-} - {}^{-}\neg)L = \neg\overline{L} - \overline{\neg L} =$$

$$= \neg(\hat{1} - L) - (\hat{1} - \neg L) = \neg\hat{1} - \neg L - (\hat{1} - \neg L) =$$

$$= (\hat{1} - \neg L) - (\hat{1} - \neg L) = 0\,.$$

Problem

Prove the commutation of the complement operator and the operator of the orthogonal transformation θ.

Solution

$$[\theta, {}^{-}] = \theta\overline{L} - \overline{\theta L} =$$

$$= \theta(\hat{1} - L) - (\hat{1} - \theta L) =$$

$$= (\hat{1} - \theta L) - (\hat{1} - \theta L) = 0\,.$$

Problem

Prove that a change of sign in the commutator of modal connectives is equivalent to the orthogonal transformation of the commutator:

$$[\Diamond, \Box] = -[\Box, \Diamond] = \theta\,[\Box, \Diamond]\,.$$

Solution

$$[\Diamond, \Box] = \Diamond\Box - \Box\Diamond\,.$$

Since

$$\Diamond = \neg\Box\neg$$

we get

$$\Diamond\Box - \Box\Diamond = \neg\Box\neg\Box - \Box\neg\Box\neg$$

Since

$$\neg\Box = \Diamond\neg \qquad \text{and} \qquad \Box\neg = \neg\Diamond$$

it follows that

$$\urcorner\Box\urcorner\Box - \Box\urcorner\Box\urcorner = \urcorner\Box\Diamond\urcorner - \urcorner\Diamond\Box\urcorner$$
$$= \urcorner(\Box\Diamond - \Diamond\Box)\urcorner =$$
$$= \theta\,[\Box, \Diamond]\,.$$

Problem

Which pairs of complementary logic operators commute?

Solution

$$[\,\mathrm{I}, \urcorner\,] = 0$$
$$[\,\hat{1}, \hat{0}\,] = 0$$

Problem

Which pairs of dual logic operators commute?

Solution

$$[\rightarrow, /] = 0\,, \qquad [\perp, \perp] = 0$$
$$[\leftarrow, \backslash] = 0\,, \qquad [\top, \top] = 0$$
$$[\,\mathrm{I}, \urcorner\,] = 0\,, \qquad [\dashv, \dashv] = 0$$
$$[\,\hat{1}, \hat{0}\,] = 0\,, \qquad [\vdash, \vdash] = 0$$

Problem

Which logic operators commute with any other logic operator?

Solution

I and $\hat{0}$.

Problem

Find all the commutator products generating the comparison operators ▼ and ▲.

Solution

24 commutators:

$$\pm[\wedge, \vee], \qquad \pm[\vee, \urcorner], \qquad \pm[\urcorner, \downarrow]$$
$$\pm[\wedge, \urcorner], \qquad \pm[\vee, \uparrow], \qquad \pm[\urcorner, \uparrow]$$
$$\pm[\wedge, \downarrow], \qquad \pm[\vee, \hat{1}], \qquad \pm[\hat{1}, \downarrow]$$
$$\pm[\wedge, \hat{1}], \qquad \pm[\downarrow, \uparrow], \qquad \pm[\hat{1}, \uparrow]$$

All the operators involved are symmetric operators.

Problem

Calculate the tautology commutator for the basic set of 16 positive-valued logic functions, the disjunction inverse $\vee^{-1}$ and the antiNAND function $-\downarrow$. Carry out the differentiation with respect to the variable p.

Solution

$$[^-,'] \ <p|\leftrightarrow|q> = \overline{2q-1} - 1 + 2q = 1$$

$$[^-,'] \ <p|\urcorner|q> = \overline{1-2q} - 2q + 1 = 1$$

$$[^-,'] \ <p|\wedge|q> = \bar{q} - (-q) = \bar{q} + q = 1$$

$$[^-,'] \ <p|\vee|q> = q + \bar{q} = 1$$

$$[^-,'] \ <p|\rightarrow|q> = \overline{-\bar{q}} - \bar{q} = 1 + \bar{q} - \bar{q} = 1$$

$$[^-,'] \ <p|\leftarrow|q> = \bar{q} + q = 1$$

$$[^-,'] \ <p|\backslash|q> = q + \bar{q} = 1$$

$$[^-,'] \ <p|/|q> = \overline{-q} - q = 1 + q - q = 1$$

$$[^-,'] \ <p|\downarrow|q> = \overline{-q} - q = 1 + q - q = 1$$

$$[^-,'] \ <p|\uparrow|q> = \overline{-\bar{q}} - \bar{q} = 1 + \bar{q} - \bar{q} = 1$$

$$[^-,'] \ <p|\perp|q> = \bar{1} + 1 = 0 + 1 = 1$$

$$[^-,'] \ <p|\mathrm{T}|q> = \overline{-1} - 1 = 1 + 1 - 1 = 1$$

$$[^-,'] \ <p|\dashv|q> = \bar{0} - 0 = 1 - 0 = 1$$

$$[^-,'] \ <p|\vdash|q> = \bar{0} - 0 = 1$$

$$[^-,'] \ <p|\hat{1}|q> = \bar{0} - 0 = 1$$

$$[^-,'] \ <p|\hat{0}|q> = \bar{0} - 0 = 1$$

$$[^-,'] \ <p|\vee^{-1}|q> = \overline{<p|\vee^{-1}|q>'_p} - <p|\overline{\vee^{-1}}|q>'_p =$$

$$= \overline{(-\bar{p}\,\bar{q} + \bar{p}\,q + p\,\bar{q})'_p} - (2\bar{p}\,\bar{q} - p\,q)'_p =$$

$$= \overline{\bar{q} - q + \bar{q}} + 2\bar{q} - q = 1$$

$$[^-,'] \ <p|-\downarrow|q> = \overline{(-\bar{p}\,\bar{q} - \bar{p}\,q - p\,\bar{q})'_p} - (2\bar{p}\,\bar{q} + 2\bar{p}\,q + 2p\,\bar{q} + p\,q)'_p =$$

$$= \bar{q} + q = 1 .$$

Problem

Of the 256 commutator products of positive logic how many are inadmissible in the discrete universe $\Omega^{\square}$, and which are they?

Solution

14 forbidden commutators:

$\pm[\wedge, \downarrow]$,
$\pm[\wedge, \mathrm{T}]$,
$\pm[\wedge, \vdash]$,
$\pm[\downarrow, \perp]$,
$\pm[\downarrow, \dashv]$,
$\pm[\perp, \vdash]$,
$\pm[\mathrm{T}, \dashv]$.

Problem

In the bracket product

$$(0\ \ 1)\begin{pmatrix} 2 & 2 \\ 0 & 2 \end{pmatrix}\begin{pmatrix} \frac{1}{2} \\ \frac{1}{2} \end{pmatrix} = 1$$

express the matrix operator as a product of two binary logic operators. Will this product be admissible in the discrete universe?

Solution

The tautology operator is equivalent to the logic quantum number 2.

$$(0\ \ 1)\begin{pmatrix} 2 & 2 \\ 0 & 2 \end{pmatrix}\begin{pmatrix} \frac{1}{2} \\ \frac{1}{2} \end{pmatrix} = (0\ \ 1)\begin{pmatrix} 1 & 1 \\ 0 & 1 \end{pmatrix} 2 \begin{pmatrix} \frac{1}{2} \\ \frac{1}{2} \end{pmatrix} =$$

$$= (0\ \ 1)\begin{pmatrix} 1 & 1 \\ 0 & 1 \end{pmatrix}\begin{pmatrix} 1 & 1 \\ 1 & 1 \end{pmatrix}\begin{pmatrix} \frac{1}{2} \\ \frac{1}{2} \end{pmatrix} =$$

$$= \langle 1 | \rightarrow \hat{1} | \tfrac{1}{2} \rangle = 1 .$$

In the discrete universe the interaction $\rightarrow \hat{1}$ is forbidden.

Problem

Consider a fermion spin lattice:

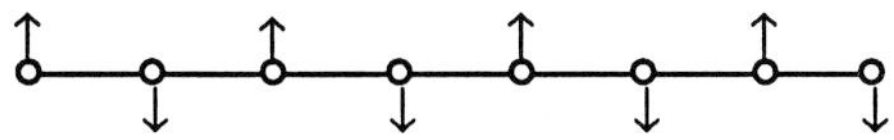

and the associated logic products of Kronecker delta pulses:

<1 | 1> <0 | 0> <1 | 1> <0 | 0>...

Obtain contraction of the products for even and odd numbers of spins.

Solution

The lattice is a tautology independently of the number of spins:

$\alpha \equiv 1$.

Even number of spins:

α = <1 |\/\/\/\| 0> = <1 |\| 0> = 1

or

α = <0 |/\/\/\/| 1> = <0 |/| 1> = 1

Odd number of spins:

α = <1 |\/\/\/| 1> = <1 | ∧ | 1> = 1

or

α = <0 |/\/\/\| 0> = <0 | ↑ | 0> = 1

Problem

The norm of which variable logic operator is invariably inversely proportional to the quark charge?

Solution

$$\frac{1}{\|\overline{\Omega}\|} = \frac{1}{\overline{\bar{p}\,\bar{q}} + \overline{\bar{p}\,q} + \overline{p\,\bar{q}} + \overline{p\,q}} \equiv \frac{1}{3}$$

The negation operator in the denominator acts as a linear operator:

$$\urcorner(\bar{p}\,\bar{q} + \bar{p}\,q + p\,\bar{q} + p\,q) = \overline{\bar{p}\,\bar{q}} + \overline{\bar{p}\,q} + \overline{p\,\bar{q}} + \overline{p\,q}$$

Problem

Show that the diagonal form of the filters $\perp$, $\top$, $\dashv$, $\vdash$ coincides with the conjunction operator $\wedge$.

Solution

Filter $\perp$:

$$\begin{pmatrix} \frac{1}{2} & 0 \\ -\frac{1}{2} & 1 \end{pmatrix}^{-1} \begin{pmatrix} 0 & 0 \\ 1 & 1 \end{pmatrix} \begin{pmatrix} \frac{1}{2} & 0 \\ -\frac{1}{2} & 1 \end{pmatrix} = \begin{pmatrix} 2 & 0 \\ 1 & 1 \end{pmatrix} \begin{pmatrix} 0 & 0 \\ 1 & 1 \end{pmatrix} \begin{pmatrix} \frac{1}{2} & 0 \\ -\frac{1}{2} & 1 \end{pmatrix} = \begin{pmatrix} 0 & 0 \\ 0 & 1 \end{pmatrix}$$

Converse filter $\dashv$:

$$\begin{pmatrix} 1 & \frac{1}{2} \\ 0 & \frac{1}{2} \end{pmatrix}^{-1} \begin{pmatrix} 0 & 1 \\ 0 & 1 \end{pmatrix} \begin{pmatrix} 1 & \frac{1}{2} \\ 0 & \frac{1}{2} \end{pmatrix} = \begin{pmatrix} 1 & -1 \\ 0 & 2 \end{pmatrix} \begin{pmatrix} 0 & 1 \\ 0 & 1 \end{pmatrix} \begin{pmatrix} 1 & \frac{1}{2} \\ 0 & \frac{1}{2} \end{pmatrix} = \begin{pmatrix} 0 & 0 \\ 0 & 1 \end{pmatrix}$$

Antifilter $\top$:

$$\begin{pmatrix} \frac{1}{2} & 1 \\ -\frac{1}{2} & 0 \end{pmatrix}^{-1} \begin{pmatrix} 1 & 1 \\ 0 & 0 \end{pmatrix} \begin{pmatrix} \frac{1}{2} & 1 \\ -\frac{1}{2} & 0 \end{pmatrix} = \begin{pmatrix} 0 & -2 \\ 1 & 1 \end{pmatrix} \begin{pmatrix} 1 & 1 \\ 0 & 0 \end{pmatrix} \begin{pmatrix} \frac{1}{2} & 1 \\ -\frac{1}{2} & 0 \end{pmatrix} = \begin{pmatrix} 0 & 0 \\ 0 & 1 \end{pmatrix}$$

Converse antifilter $\vdash$:

$$\begin{pmatrix} 0 & \frac{1}{2} \\ 1 & \frac{1}{2} \end{pmatrix}^{-1} \begin{pmatrix} 1 & 0 \\ 1 & 0 \end{pmatrix} \begin{pmatrix} 0 & \frac{1}{2} \\ 1 & \frac{1}{2} \end{pmatrix} = \begin{pmatrix} -1 & 1 \\ 2 & 0 \end{pmatrix} \begin{pmatrix} 1 & 0 \\ 1 & 0 \end{pmatrix} \begin{pmatrix} 0 & \frac{1}{2} \\ 1 & \frac{1}{2} \end{pmatrix} = \begin{pmatrix} 0 & 0 \\ 0 & 1 \end{pmatrix}$$

Problem

Employing the eigenequation for the negation operator, eliminate the operator in the following products:

$$\text{(a)}\ \neg\begin{pmatrix}0\\1\end{pmatrix},\quad \text{(b)}\ \neg\begin{pmatrix}\frac{1}{4}\\\frac{3}{4}\end{pmatrix},\quad \text{(c)}\ \neg\begin{pmatrix}\frac{2}{5}\\\frac{3}{5}\end{pmatrix}.$$

Solution

(a)
$$\neg\begin{pmatrix}0\\1\end{pmatrix} = \neg\left[\begin{pmatrix}\frac{1}{2}\\\frac{1}{2}\end{pmatrix} + (-1)\begin{pmatrix}\frac{1}{2}\\-\frac{1}{2}\end{pmatrix}\right] = \neg\begin{pmatrix}\frac{1}{2}\\\frac{1}{2}\end{pmatrix} - \neg\begin{pmatrix}\frac{1}{2}\\-\frac{1}{2}\end{pmatrix} =$$
$$= 1\begin{pmatrix}\frac{1}{2}\\\frac{1}{2}\end{pmatrix} - (-1)\begin{pmatrix}\frac{1}{2}\\-\frac{1}{2}\end{pmatrix} =$$
$$= \begin{pmatrix}\frac{1}{2}\\\frac{1}{2}\end{pmatrix} + \begin{pmatrix}\frac{1}{2}\\-\frac{1}{2}\end{pmatrix} = \begin{pmatrix}1\\0\end{pmatrix}.$$

(b)
$$\neg\begin{pmatrix}\frac{1}{4}\\\frac{3}{4}\end{pmatrix} = \neg\left[\begin{pmatrix}\frac{1}{2}\\\frac{1}{2}\end{pmatrix} + (-\tfrac{1}{2})\begin{pmatrix}\frac{1}{2}\\-\frac{1}{2}\end{pmatrix}\right] = \neg\begin{pmatrix}\frac{1}{2}\\\frac{1}{2}\end{pmatrix} - \frac{1}{2}\neg\begin{pmatrix}\frac{1}{2}\\-\frac{1}{2}\end{pmatrix} =$$
$$= 1\begin{pmatrix}\frac{1}{2}\\\frac{1}{2}\end{pmatrix} - \frac{1}{2}(-1)\begin{pmatrix}\frac{1}{2}\\-\frac{1}{2}\end{pmatrix} =$$
$$= \begin{pmatrix}\frac{1}{2}\\\frac{1}{2}\end{pmatrix} + \frac{1}{2}\begin{pmatrix}\frac{1}{2}\\-\frac{1}{2}\end{pmatrix} = \begin{pmatrix}\frac{3}{4}\\\frac{1}{4}\end{pmatrix}.$$

(c)
$$\neg\begin{pmatrix}\frac{2}{5}\\\frac{3}{5}\end{pmatrix} = \neg\left[\begin{pmatrix}\frac{1}{2}\\\frac{1}{2}\end{pmatrix} + (-\tfrac{1}{5})\begin{pmatrix}\frac{1}{2}\\-\frac{1}{2}\end{pmatrix}\right] = \neg\begin{pmatrix}\frac{1}{2}\\\frac{1}{2}\end{pmatrix} - \frac{1}{5}\neg\begin{pmatrix}\frac{1}{2}\\-\frac{1}{2}\end{pmatrix} =$$
$$= 1\begin{pmatrix}\frac{1}{2}\\\frac{1}{2}\end{pmatrix} - \frac{1}{5}(-1)\begin{pmatrix}\frac{1}{2}\\-\frac{1}{2}\end{pmatrix} =$$
$$= \begin{pmatrix}\frac{1}{2}\\\frac{1}{2}\end{pmatrix} + \frac{1}{5}\begin{pmatrix}\frac{1}{2}\\-\frac{1}{2}\end{pmatrix} = \begin{pmatrix}\frac{3}{5}\\\frac{2}{5}\end{pmatrix}.$$

Problem

An operator L' acting on the eigenstate vector $|q>$ can decay into the sum of the operators L", L'''

$$L'|q> \; = \; (L'' + L''')|q> \; = \; L''|q> + L'''|q>$$

on condition that L' = L" + L''' and that $|q>$ is the eigenvector for L" and L'''. Then $\lambda' = \lambda'' + \lambda'''$.
Find the set of elementary logic operators of which further decay is not possible.

Solution

Operator decay	Eigenstate	$\lambda = \Sigma\, \lambda_i$
$I = \uparrow + \wedge$	any	$1 = 0\,(1) + 1\,(0)$
$\rightarrow = \uparrow + / + \wedge$ $\rightarrow = I + /$ $\rightarrow = \mathrm{T} + \wedge$ $\rightarrow = \uparrow + \dashv$	$\begin{pmatrix}1\\0\end{pmatrix}$	$1 = 1 + 0 + 0$ $1 = 1 + 0$ $1 = 1 + 0$ $1 = 1 + 0$
$\leftarrow = \uparrow + \backslash + \wedge$ $\leftarrow = I + \backslash$ $\leftarrow = \vdash + \wedge$ $\leftarrow = \uparrow + \perp$	$\begin{pmatrix}0\\1\end{pmatrix}$	$1 = 0 + 0 + 1$ $1 = 1 + 0$ $1 = 0 + 1$ $1 = 0 + 1$
$\perp = \backslash + \wedge$ $\mathrm{T} = \uparrow + /$ $\dashv = / + \wedge$ $\vdash = \uparrow + \backslash$	$\|1>$ $\|0>$ $\|0>$ $\|1>$	$1 = 0 + 1$ $1 = 1 + 0$ $0 = 0 + 0$ $0 = 0 + 0$
$\hat{1} = I + \urcorner$	$\begin{pmatrix}\frac{1}{2}\\\frac{1}{2}\end{pmatrix}, \begin{pmatrix}\frac{1}{2}\\-\frac{1}{2}\end{pmatrix}$	$2 = 1 + 1$ $0 = 1 - 1$

Only binary eigenvectors can be considered for decay of the filters ⊥, T, ⊣, ⊢, because ∧, \, /, ↑ have only binary eigenstates. For tautology decay, further decay of the negation and identity operators is not possible, since $\hat{1}$ does not share the same eigenstates with \ and /, and ↑ and ∧ respectively. Thus, the set of elementary logic operators:

$$\{ I, \urcorner, \wedge, \backslash, /, \uparrow, \hat{0} \} .$$

Problem

Employing the operator base of elementary operators:

$$\{ \leftrightarrow, \urcorner, \wedge, \backslash, /, \uparrow, \hat{0} \} ,$$

construct by juxtaposition the 4-D hyperspace hyperconnectives H which

(a) are orthogonal operators,

$$\tilde{H} = H^{-1}$$

$$<h_i \mid H \mid h_j> = <h_j \mid H^{-1} \mid h_i> ,$$

$$<h_i \mid H\tilde{H} \mid h_j> = <h_j \mid h_i> ,$$

(b) form a group with repsect to the operation of matrix multiplication.

Solution

$$\begin{pmatrix} \leftrightarrow & 0 \\ 0 & \leftrightarrow \end{pmatrix} \quad \begin{pmatrix} \leftrightarrow & 0 \\ 0 & \urcorner \end{pmatrix} \quad \begin{pmatrix} \urcorner & 0 \\ 0 & \leftrightarrow \end{pmatrix} \quad \begin{pmatrix} \uparrow & \wedge \\ \wedge & \uparrow \end{pmatrix}$$

$$\begin{pmatrix} \wedge & \uparrow \\ \uparrow & \wedge \end{pmatrix} \quad \begin{pmatrix} \wedge & / \\ \backslash & \uparrow \end{pmatrix} \quad \begin{pmatrix} \uparrow & \backslash \\ / & \wedge \end{pmatrix} \quad \begin{pmatrix} \urcorner & 0 \\ 0 & \urcorner \end{pmatrix}$$

$$\begin{pmatrix} 0 & \leftrightarrow \\ \leftrightarrow & 0 \end{pmatrix} \quad \begin{pmatrix} 0 & \urcorner \\ \urcorner & 0 \end{pmatrix} \quad \begin{pmatrix} / & \backslash \\ \backslash & / \end{pmatrix} \quad \begin{pmatrix} \backslash & / \\ / & \backslash \end{pmatrix}$$

$$\begin{pmatrix} \wedge & / \\ \uparrow & \backslash \end{pmatrix} \quad \begin{pmatrix} \wedge & \uparrow \\ \backslash & / \end{pmatrix} \quad \begin{pmatrix} \uparrow & \backslash \\ \wedge & / \end{pmatrix} \quad \begin{pmatrix} \uparrow & \wedge \\ / & \backslash \end{pmatrix}$$

$$\begin{pmatrix} / & \wedge \\ \backslash & \uparrow \end{pmatrix} \quad \begin{pmatrix} \backslash & / \\ \wedge & \uparrow \end{pmatrix} \quad \begin{pmatrix} / & \backslash \\ \uparrow & \wedge \end{pmatrix} \quad \begin{pmatrix} \backslash & \uparrow \\ / & \wedge \end{pmatrix}$$

$$\begin{pmatrix} 0 & \leftrightarrow \\ \rceil & 0 \end{pmatrix} \quad \begin{pmatrix} 0 & \rceil \\ \leftrightarrow & 0 \end{pmatrix} \quad \begin{pmatrix} / & \wedge \\ \uparrow & \backslash \end{pmatrix} \quad \begin{pmatrix} \backslash & \uparrow \\ \wedge & / \end{pmatrix}$$

How many different group logic operators H can be constructed in an n-dimensional logic space?

Solution

$n!$

Employ the transition from left to right logic to solve the inverse problem for group hyperconnectives

$$| h_j \rangle = H \, | h_i \rangle .$$

Solution

$$\widetilde{| h_j \rangle} = \widetilde{H | h_i \rangle}$$

$$\widetilde{| h_i \rangle} \widetilde{H} = \widetilde{| h_j \rangle}$$

$$\langle h_i | H^{-1} = \langle h_j |$$

$$\langle h_i | H^{-1} H = \langle h_j | H$$

$$\langle h_i | = \langle h_j | H$$

In group logic the solution of the inverse problem is achieved by a simple interchange of transposed inputs and outputs, with no modification of a group operator being required. Feed back the transposed output to obtain its counterimage.

Problem

Show that the general covariancy rule

$$L = \theta \overset{\sim}{\overset{\urcorner}{L}}$$

represents the identity of the rectangular group. Describe the group in terms of diagonal and rectangular matrix rotations.

Solution

$$\overset{\sim}{\overset{\urcorner}{L}} = \theta L$$

*	I	$\sim$	$\urcorner$	θ
I	I	$\sim$	$\urcorner$	θ
$\sim$	$\sim$	I	θ	$\urcorner$
$\urcorner$	$\urcorner$	θ	I	$\sim$
θ	θ	$\urcorner$	$\sim$	I

I – identity
$\sim$ – transposition
$\urcorner$ – converse transposition
θ – orthogonal transform

$$\overset{RC}{L} = \overset{\pi}{L}$$

*	I	R	C	π
I	I	R	C	π
R	R	I	π	C
C	C	π	I	R
π	π	C	R	I

R – row transposition
C – column transposition
π – inversion

$$\overset{\sim}{\overset{\urcorner}{L}} = \overset{RC}{L} = \overset{\pi}{L} = \theta L$$

Problem

Which compound logic operators of continuous logic can be generated through anti-commutation of their constituent operators?

Solution

$$\wedge + \vee = \{\wedge, \vee\}$$
$$\uparrow + \downarrow = \{\uparrow, \downarrow\}$$

Establish the special rule for the passage from these anticommutators to the corresponding commutators.

Solution

$$\{\wedge, \vee\} = \overline{\rceil[\wedge, \vee]}$$
$$\{\uparrow, \downarrow\} = \overline{\rceil[\uparrow, \downarrow]}$$

Problem

Employ the commutator and anticommutator for conjunction and disjunction to obtain duals for compound operators:

$$(\wedge + \vee)\,,\ (/ + \rightarrow)\,,\ (\backslash + \leftarrow)\,,\ (\uparrow + \downarrow)$$

Solution

$$(\wedge + \vee)^* = \theta\overline{\begin{pmatrix} 0 & 1 \\ 1 & 2 \end{pmatrix}} = \begin{pmatrix} -1 & 0 \\ 0 & 1 \end{pmatrix} = \overline{\{\uparrow, \downarrow\}}$$

$$(/ + \rightarrow)^* = \theta\overline{\begin{pmatrix} 1 & 2 \\ 0 & 1 \end{pmatrix}} = \blacktriangledown = [\wedge, \vee]$$

$$(\backslash + \leftarrow)^* = \theta\overline{\begin{pmatrix} 1 & 0 \\ 2 & 1 \end{pmatrix}} = \blacktriangle = [\uparrow, \downarrow]$$

$$(\uparrow + \downarrow)^* = \theta\overline{\begin{pmatrix} 2 & 1 \\ 1 & 0 \end{pmatrix}} = \begin{pmatrix} 1 & 0 \\ 0 & -1 \end{pmatrix} = \overline{\{\wedge, \vee\}}$$

Problem

Pass from the scalar to the matrix code by proving the equalities:

$$p + q = < p\begin{pmatrix} 0 & 1 \\ 1 & 2 \end{pmatrix} q >$$

$$\bar{p} + q = \langle p \begin{pmatrix} 1 & 2 \\ 0 & 1 \end{pmatrix} q \rangle$$

$$p + \bar{q} = \langle p \begin{pmatrix} 1 & 0 \\ 2 & 1 \end{pmatrix} q \rangle$$

$$\bar{p} + \bar{q} = \langle p \begin{pmatrix} 2 & 1 \\ 1 & 0 \end{pmatrix} q \rangle$$

Solution

$$\bar{p} + q = \langle p | 0 \rangle + \langle 1 | q \rangle = \langle p | \top | q \rangle + \langle p | \dashv | q \rangle = \langle p | \top + \dashv | q \rangle =$$

$$= \langle p \begin{pmatrix} 1 & 2 \\ 0 & 1 \end{pmatrix} q \rangle$$

$$p + \bar{q} = \langle p | 1 \rangle + \langle 0 | q \rangle = \langle p | \perp | q \rangle + \langle p | \vdash | q \rangle = \langle p | \perp + \vdash | q \rangle =$$

$$= \langle p \begin{pmatrix} 1 & 0 \\ 2 & 1 \end{pmatrix} q \rangle$$

The solutions for $p + q$ and $\bar{p} + \bar{q}$ are similar.

Problem

Two logic vectors generate (2 × 2) logic operators,

$$| x \rangle \otimes \langle y | = L .$$

How many vectors are required for the generation of (4 × 4) hyperspace logic operators? Construct the corresponding tensor products.

Solution

Four logic vectors are required for the generation of 4-D logic operators:

$$\langle x | \otimes | y \rangle \otimes \langle z | \otimes | t \rangle = (4 \times 4) \text{ operator}$$

$$\langle x | \otimes \langle y | \otimes | z \rangle \otimes | t \rangle = (4 \times 4) \text{ operator}$$

$$| x \rangle \otimes | y \rangle \otimes \langle z | \otimes \langle t | = (4 \times 4) \text{ operator}$$

$$| x \rangle \otimes \langle y | \otimes | z \rangle \otimes \langle t | = (4 \times 4) \text{ operator}$$

Analyse the difference and the Hemming distances between the operators.

Problem

Employing the dimensional regression rule of tensor logic, evaluate the inverse for the 4-D logic operator

$$H = \begin{pmatrix} \leftarrow & \leftarrow \\ \hat{0} & \leftarrow \end{pmatrix}$$

Solution

$$H^{-1} = \begin{pmatrix} \leftarrow & \leftarrow \\ \hat{0} & \leftarrow \end{pmatrix}^{-1} = (\rightarrow \otimes \leftarrow)^{-1} =$$

$$= \rightarrow^{-1} \otimes \leftarrow^{-1} = \begin{pmatrix} 1 & 1 \\ 0 & 1 \end{pmatrix}^{-1} \otimes \begin{pmatrix} 1 & 0 \\ 1 & 1 \end{pmatrix}^{-1} =$$

$$= \begin{pmatrix} 1 & -1 \\ 0 & 1 \end{pmatrix} \otimes \begin{pmatrix} 1 & 0 \\ -1 & 1 \end{pmatrix} = \begin{pmatrix} \leftarrow^{-1} & -\leftarrow^{-1} \\ \hat{0} & \leftarrow^{-1} \end{pmatrix}.$$

$$HH^{-1} = \begin{pmatrix} \leftarrow & \leftarrow \\ \hat{0} & \leftarrow \end{pmatrix} \begin{pmatrix} \leftarrow^{-1} & -\leftarrow^{-1} \\ \hat{0} & \leftarrow^{-1} \end{pmatrix} = \begin{pmatrix} I & -I+I \\ \hat{0} & I \end{pmatrix} = I_4 .$$

Problem

Prove that with an inversion of logic space the antifractal contradiction product

$$<x_1\, x_2\, x_3\, x_4 \mid \rightarrow \otimes \backslash \mid y_1\, y_2\, y_3\, y_4>$$

retains antifractal architecture.

Solution

$$<x_1\, x_2\, x_3\, x_4 \mid F_1 \mid y_1\, y_2\, y_3\, y_4> =$$

$$= <x_1\, x_2\, x_3\, x_4 \mid \rightarrow \otimes \backslash \mid y_1\, y_2\, y_3\, y_4> =$$

$$= <x_4\, x_3\, x_2\, x_1 \mid \theta\, (\rightarrow \otimes \backslash) \mid y_4\, y_3\, y_2\, y_1> =$$

$$= <x_4\, x_3\, x_2\, x_1 \mid \theta \rightarrow \otimes\, \theta \backslash \mid y_4\, y_3\, y_2\, y_1> =$$

$$= <x_4\, x_3\, x_2\, x_1 \mid \leftarrow \otimes / \mid y_4\, y_3\, y_2\, y_1> =$$

$$= <x_4\, x_3\, x_2\, x_1 \mid F_2 \mid y_4\, y_3\, y_2\, y_1> .$$

Prove that this is a general property of antifractal logic operators. Show that the antifractal architecture of an operator is also retained with the transposition of the hyperspace bracket product:

$$<h_i \mid F \mid h_j> \; = \; <h_j \mid \tilde{F} \mid h_i> .$$

Problem

In 2-D logic space the identity and negation operators can be generated as transmultiplication products of the set { →, ←, v, ↓ }:

$$I_2 = \rightarrow \ast \leftarrow = \begin{pmatrix} 1 & 0 \\ 0 & 1 \end{pmatrix}$$

$$\neg_2 = \vee \ast \leftarrow = \begin{pmatrix} 0 & 1 \\ 1 & 0 \end{pmatrix}$$

Employing the same set of operators, construct a rule for generating the identity and negation operators in 4-D logic space.

Solution

$$I_4 = (\rightarrow \otimes \leftarrow) \ast (\leftarrow \otimes \rightarrow) = \begin{pmatrix} 1 & 0 & 0 & 0 \\ 0 & 1 & 0 & 0 \\ 0 & 0 & 1 & 0 \\ 0 & 0 & 0 & 1 \end{pmatrix}$$

$$\neg_4 = (\vee \otimes \downarrow) \ast (\downarrow \otimes \vee) = \begin{pmatrix} 0 & 0 & 0 & 1 \\ 0 & 0 & 1 & 0 \\ 0 & 1 & 0 & 0 \\ 1 & 0 & 0 & 0 \end{pmatrix}$$

Show that the rule can be employed for generating 4-D counterparts of all 16 basis logic operators, except the generators of identity and negation { →, ←, v, ↓ }.
Why?

Problem

Which hyperspace logic operator does the following truth matrix represent? Write down the corresponding fundamental product.

$$H = \begin{pmatrix}
1&1&1&1&1&1&1&1&1&1&1&1&1&1&1&1\\
0&1&0&1&0&1&0&1&0&1&0&1&0&1&0&1\\
0&0&1&1&0&0&1&1&0&0&1&1&0&0&1&1\\
0&0&0&1&0&0&0&1&0&0&0&1&0&0&0&1\\
0&0&0&0&1&1&1&1&0&0&0&0&1&1&1&1\\
0&0&0&0&0&1&0&1&0&0&0&0&0&1&0&1\\
0&0&0&0&0&0&1&1&0&0&0&0&0&0&1&1\\
0&0&0&0&0&0&0&1&0&0&0&0&0&0&0&1\\
0&0&0&0&0&0&0&0&1&1&1&1&1&1&1&1\\
0&0&0&0&0&0&0&0&0&1&0&1&0&1&0&1\\
0&0&0&0&0&0&0&0&0&0&1&1&0&0&1&1\\
0&0&0&0&0&0&0&0&0&0&0&1&0&0&0&1\\
0&0&0&0&0&0&0&0&0&0&0&0&1&1&1&1\\
0&0&0&0&0&0&0&0&0&0&0&0&0&1&0&1\\
0&0&0&0&0&0&0&0&0&0&0&0&0&0&1&1\\
0&0&0&0&0&0&0&0&0&0&0&0&0&0&0&1
\end{pmatrix}$$

Solution

The hyperspace logic operator in question is fractal implication $H = \rightarrow^{[4]}$.

$$\langle h_i \,|\rightarrow^{[4]}|\, h_j\rangle \;=\; \langle x \,|\rightarrow|\, y\rangle\, \langle x' \,|\rightarrow|\, y'\rangle\, \langle x'' \,|\rightarrow|\, y''\rangle\, \langle x''' \,|\rightarrow|\, y'''\rangle\,.$$

INDEX

Abstraction 56, 165
Action quantum 207
Affirmation 6, 150
Annihilation interactions 58, 256
Annihilation operator 207
Antecedent 10, 126
Anticommutation 211, 216, 282
Antifilter 14, 43
Antifractal 235, 284
Antilogic 137
Antiparticle 197, 200
Artificial intelligence 160
Atomic product 47
Autoproduct 164

Baryon octet 200
Basis set 17, 75
    major 18, 60, 75
    minimal 19
    minor 18, 75
    standard 18
Binary
    connective 8
    operator 49
Bohr 195
Boole 5
Boolean
    algebra 20, 80
    connectives 6, 193
    functions 12
    logic 5, 49, 109
Bracket 22
    product 47, 121
Brain 197, 229, 240
Bra-vector 37
Boson 209

Cancellation rule 57
Charge operator 224, 226
Chronologic 156
Closure relation 39, 208
Codeword 223
Cognition 3, 89, 96, 177, 204
Commutation relation 59, 187
Comparison operator 154, 218
Complement operator 61, 115
Conjunction 8, 27, 31, 43, 80, 101, 276
Conjunctive normal form 24
Connectives 6, 8, 29, 106, 140, 229
    interaction 2, 56, 81
Consequent 10, 25, 126
Conservation laws 96, 202
Continuous universe 97, 162
Contradiction 85, 91
    law 23, 109, 137
    subtraction 84, 88
Contraposition 44, 69
Converse
    logic function 15, 127
    transposition 44, 69
Covariance 3, 72
Creation operator 207
Creativity 165, 205

De Morgan laws 18, 23, 64, 79, 105
Deterministic logic 5, 46, 98, 101

Diagonalization 193, 276
Differential logic 122
Dirac 152, 219
- bispinor 219, 226
- brackets 37
- spinor matrices 219

Discrete universe 46
Disjunction 9, 27, 33, 43, 101, 181
Disjunctive normal form 24
Distributive laws 5, 23
Dual
- basis 64
- connective 22, 65
- vector space 38

Duality principle 22, 64

Eigenequation 118, 175
Eigenvalue 3, 119, 175
- spectrum 3, 186

Eigenvector 175
Electron 152, 220
Empty operator 43
Equivalence 11, 49, 167
Excluded Middle law 23, 26, 92
Exclusion principle 220
Exteriorization 167
Extremum 134

Falsum 38
Fermion 211, 275
Filter 14, 43, 255
Forbidden interactions 2, 56, 161
Fractal
- operator 234
- quantum mechanics 195

Fundamental
- interactions 197
- product 80, 121,145, 191, 234

Group
- linear transformations 219
- logic 279
- Lorentz 219
- operator 280
- symmetry 215

Hamiltonian 177
Hemming 281
Hilbert 37, 173
High-level intelligence 89, 195
Hyperconnective 229
Hypercube 201
Hyperspace
- operator 231
- vector 231

Idempotence 21, 78, 84, 109, 178
Identity 6, 71
Inhibition 43
Integration constants 132
Imaginary truth-value 259
Implication 10, 43, 103, 126
- algebra 85
- converse

Inference rules 89
Inner product 46, 121
Intellect 91, 120, 165
Intelligence code 3, 89, 135, 165
Interiorization 167
Inverse problem 280
Isospin 223
Isotropic interval 242

Juxtaposition 81, 231, 236, 279

Ket-vector 37
Kronecker function 40, 46, 81, 83

Left logic 50, 63, 103, 188
Linear
independence 39, 75
operator 237
Logic
addition 22, 264
charge 226
constants 7
coordinates 41, 54, 143, 241
derivative 122
generators 76
matrix base 74, 95, 220
norm 54, 200
operator 42, 77
quantum numbers 3, 177
space 37
subspace 236
vector 37

Lorentz 201, 219

Matrix
antisymmetric 154
binary 43
bi-asymmetric 45, 71
bisymmetric 45, 71
contrasymmetric 45, 70
diagonal 44
idempotent 254
inverse 3, 138
null 58
orthogonal 64, 279
partitioned 232
resolvent 175
singular 139, 175
square 42, 232
string 79
symmetric 45, 69
transpose 43
unit 138, 260
Memory 165
trace 173
Memory processor 166, 173
Meson octet 200
Minkowski 37
Modal
connectives 29, 271
logic 5
negation 31
Modus Ponens 95
Modus Tollens 96
Multiplication tables 57, 60, 106, 145, 162, 192, 230, 264
Multivalued logic 5, 26

Nand 16, 19, 43, 181
Negation 7, 30
Boolean 7
cyclic 26
diametrical 26
n-D 236
operator 49, 179, 194
Negative logic 137, 162
Neutron 223
Non-commuting algebra 59, 196
Nor 16, 19, 43
Normal product 198
Normalization 29, 38, 142
Nucleon 223

Observation 59, 175
Operator
annihilation 207, 210
creation 207, 210
Hermitian 175
norm 54

spectrum 3
Orthogonal transformation 64, 72
Orthomodularity 250
Outer product 76, 111, 121

Parallel architecture 229
Pauli 211, 215, 220
Piaget 8
Positron 152
Post 27
Probability
addition law 33
conditional 35
logic 29, 101
multiplication law 32
theory 29, 98
Proton 223
Pseudo-Euclidian metric 242

Quantum logic 5, 196
Quantum theory 3, 174, 195
Quark 204, 223, 276

Relativity 37, 160
Right logic 50, 63, 103, 188
Russell paradox 263

Scalar limit 1, 164
Scalarization 2, 87
Shiefer 68
Spacetime
diagram 3, 197
Spin 215
Spinor space 219
State vector 174
Strong interactions 204, 225
Subnegators 147, 260
Subspace 240
Superassociativity 82
Symmetry 1, 45, 67, 201, 215

Tautology 23, 43, 90, 104, 180, 194
Tensor 1, 81, 229
Ternary
conjunction 28
disjunction 28
logic 27
Trace 114, 269
Transmultiplication 59, 93, 110
Transposition 44, 68
Truth-tables 6, 13, 231
Truth-value
alphabet 3, 109, 152, 164

Unary
connective 6
operator 49, 102
Uncertainty principle 160, 205
Undecided vectors 46, 51, 112
Unitary symmetry 199, 215
Universum operator 111, 268

Vacuum 151, 204
Vector
logic 77
space 39
Vectorization 2, 48, 206
Verum 38
Virtual interactions 204

Wave function 174
Wave-particle dualism 166
Wigner 211

Xor 9

Zellweger 201